Caribbean Primary

Mathematics

Level 6

6th edition

Contributors

Jonella Giffard

Hyacinth Dorleon

Melka Daniel

Martiniana Smith

Troy Nestor

Lydon Richardson

Rachel Mason

Eugenia Charles

Sharon Henry-Phillip

Glenroy Phillip

Jeffrey Blaize

Clyde Fitzpatrick

Reynold Francis

Wilma Alexander

Rodney Julien

C. Ellsworth Diamond

Shara Quinn

HODDER EDUCATION

AN HACHETTE UK COMPANY

Contents

How to use this book

This Student's Book meets the objectives of the OECS and regional curricula for Level 6. The Student's Book provides both learning notes marked as [Explain] and a wide range of activities to help and encourage students to meet the learning outcomes for the level.

The content is arranged in topics, which correspond to the strands in the syllabus. For example, Topics 2, 5 and 9 relate to the strand of Number sense, while Topics 4 and 12 relate to the various aspects that need to be covered in the strand Geometry (Shape and space).

Topic 1, **Getting ready**, is a revision topic that allows you to do a baseline assessment of key skills and concepts covered in the previous level. You may need to revise some of these concepts if students are uncertain or struggle with them.

Each topic begins with an opening spread with the following features:

Notes for teachers about key mathematical skills that need to be developed

Questions to discuss with the students as they begin a new topic

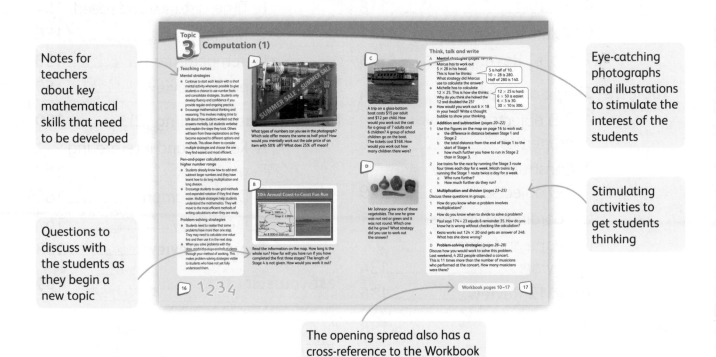

Eye-catching photographs and illustrations to stimulate the interest of the students

Stimulating activities to get students thinking

The opening spread also has a cross-reference to the Workbook pages which support the topic.

The questions and photographs relate to the specific units of each topic. You can let the students do all the activities (A, B and C, for example) as you start a topic, or you can do only the activities that relate to the unit you are about to start.

Topics 2 to 15 are sub-divided into units, which deal with different skills that need to be developed to meet the learning outcomes. For example, Topic 2, **Number sense (1)**, is divided into three units: **A** Counting and place value, **B** Rounding and estimating, and **C** Large numbers in real life.

Each unit is structured in a similar way.

A student-friendly list of learning objectives

A key word list – the words are bold and blue in the text

Graded activities for consolidation

Teaching text that explains concepts and provides examples

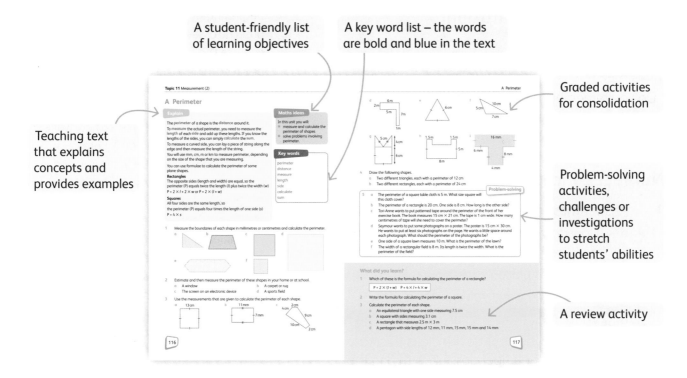

Problem-solving activities, challenges or investigations to stretch students' abilities

A review activity

As you work through the units you will find a range of different types of activities and tasks, including practical investigations, problem-solving strategies, projects and challenge questions. These features are clearly marked in the book so you know what you are dealing with.

The topics end with a review page that provides:

✳ An interactive summary called *Key ideas and concepts* to help students consolidate and reflect on what they've learnt.
✳ *Think, talk, write …* activities, which encourage students to share ideas, clarify their thinking and write in their maths journals.
✳ *Quick check* revision questions that cover all the units.

There are three tests provided in the Student's Book to allow for ongoing assessment and to prepare students for formal testing at all levels. Test 1 covers work from Topics 1 to 6, Test 2 covers work from Topics 1 to 12 and Test 3 covers work from the entire book.

Topic 1 Getting ready

Number and computation

Explain

Do you remember what you learnt in Level 5? Before you start on new work, you will revise some of the mathematics you already know.

Number concepts

1   Answer these questions about the number 41 539.

    a   How do you say this number? Write it in words.

    b   What is the face value of the digit in the thousands place?

    c   What is the place value of the digit 4?

    d   Which digit has the smallest value in this number?

    e   How can you arrange the digits to create the largest number possible with 3 in the thousands place?

    f   What number is fifty-five thousand greater than this one?

    g   How do you write the number you made in part **f** in expanded notation? Try to show two different methods.

2   Che is buying a computer that costs $2 856. He pays with $100 bills. How many notes will he give the salesperson?

3   a   Find the lowest common multiple (LCM) of 4 and 5.

    b   Find the highest common factor (HCF) of 18 and 27.

4   Draw a diagram to show a fraction greater than 1, but smaller than 2.

5   Use these fractions: $\frac{4}{6}$ $\frac{1}{4}$ $\frac{6}{5}$ $1\frac{2}{3}$

    a   Write the improper fraction as a mixed number and an equivalent decimal.

    b   Write the mixed number as an improper fraction.

    c   Which fraction is equivalent to 0.25?

    d   Write the fraction that has a denominator of 6 in simplest form.

Teaching notes

This topic revisits some of the key skills taught in Level 5.

You can use these pages at the beginning of the year, as part of your baseline assessment of students, or you can use the relevant sections as you begin each new topic with your class. If your students struggle with an activity, you may need to revise the concepts and skills before moving on to new work.

* Can students read and write numbers to 99 999 in figures and words? If not, revise and reinforce the concepts.

* Do students understand place value? If not, continue to work with place value tables to break down numbers using place value.

* Can students identify and compare fractions, including mixed numbers and improper fractions? Are they able to find equivalent fractions, decimals and percentages? Encourage students to refer to a fraction wall as needed. Students could make and colour their own one.

* Can students list factors and multiples of numbers and use them to find the HCF and LCM of sets of two or more numbers? Use factor trees, multiplication tables and skip counting to revise these ideas.

* Can students identify odd, even, prime, composite and square numbers? Revise the vocabulary and these concepts if needed.

* Do students know the basic number facts? Let them take turns to quiz and test each other on these.

* Can students use different strategies to recall multiplication and related division facts? Let students explain their strategies to each other and use them to revise and memorise the basic facts.

6 Which of the quantities in the box are equivalent to the shaded section of the diagram?

0.25	$\frac{6}{8}$	0.75	$\frac{3}{5}$	75%
50%	$\frac{4}{8}$	$\frac{12}{15}$	$\frac{12}{16}$	$\frac{1}{4}$

✳ Can students do written calculations in all four operations with larger numbers? Let them check each other's work and observe them as they do calculations to correct any mistakes. Encourage the use of different strategies.

✳ Can students calculate efficiently with fractions and decimals? Continue to use concrete models and diagrams to demonstrate and model these concepts.

7 There are 18 boys in class of 30 students. What is the ratio of boys to girls in this class?

8 Rewrite these fractions in ascending order:
$\frac{2}{3}$ $\frac{1}{2}$ $\frac{5}{6}$ $\frac{4}{9}$ $\frac{1}{6}$

9 a List the first five prime numbers.

b Rewrite the list using Roman numerals.

10 Do these calculations using written methods.

a Find the sum of 20 156, 9 046 and 198.   b What must be added to 346 to make 1 000?

c Find the product of 34 and 7.   d What is the quotient if 714 is divided by 3?

11 Marcus bought 3 items at the stationery store. He paid with $20 and received $4.60 change.

a What three items did he buy?

b What was the total cost of the items?

c The price of notebooks is increased by 10%. What is the new price?

d In a sale, you get 20% off the price of erasers and sharpeners. What would you pay for each item on sale?

12 Shaunte bakes cookies for the school fair. She has 25 baking trays. Each tray holds 1 dozen cookies.

a If she fills all the trays, how many cookies does she make?

b She makes packs of 3 cookies. How many packs can she make?

13 Calculate.

a $\frac{1}{4} + \frac{2}{3}$ b $\frac{7}{8} - \frac{1}{2}$ c $2\frac{2}{5} + 1\frac{1}{3}$

d $1\frac{1}{3} \times 1\frac{1}{2}$ e 0.2×3.5 f $3.6 \div 4$

14 Samantha gave away $\frac{5}{8}$ of a cake. How much cake does she have left?

15 Karlie has 2 kilograms of grapes. She wants to make $\frac{1}{4}$ kg packets to share among her friends. How many packets can she make?

16 Kira earns $5 for each tree she trims. She trims 6 trees each hour, 5 hours a day. After trimming trees for 2 days, how much money will Kira have earned?

17 Choose numbers from the box. Write two word problems using the numbers. Each problem should involve at least two steps and two different operations.

23 450	2.5	160	$\frac{1}{4}$	1 320	96	5	400

Geometry, measurement and data

1   Identify the shapes or objects as accurately as possible.

   a   I am a quadrilateral with equal sides and all my angles are right angles.

   b   I have six congruent faces, twelve edges and eight vertices.

   c   I have three sides, but none of them are equal.

   d   My angles add up to 360 degrees and only two of my sides are parallel.

   e   I am a plane shape. I have five straight sides.

   f   Look at these nets. What solid would you get if you folded each one up?

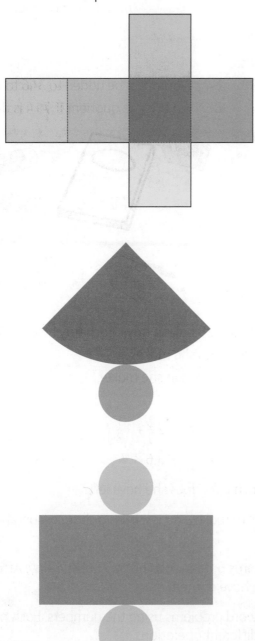

Teaching notes

*   Can students classify 2-D and 3-D shapes using shape properties? Remind students to count sides and to compare the lengths of sides and the sizes of angles. Ask questions like: How many shapes with four sides can you see? Which of these have all their sides the same length? Which shapes have only three sides? How many faces does this solid have? What shape are the faces?

*   Can students measure and name angles? Display the angle types in the classroom and allow students to measure and check to decide what type of angle they are dealing with.

*   Do students understand the concepts of perimeter and area? Can they measure and calculate perimeter and area, and give the answers in appropriate units?

*   Can students tell time up to one-minute intervals? Display a clock in the classroom or have a smaller clock for each group to use. Stop during the day at appropriate times to let them work out what time it is.

*   Are students able to work with different units of measurement and to state the relationship between these? Revise vocabulary as needed and continue to involve students in practical measuring tasks.

2   Which measuring instrument could you use to measure in these units?

| metres   litres   grams   seconds   degrees Celsius |

3   Which unit of length is best for measuring:
    a   the length of a car?
    b   the perimeter of an island?
    c   the thickness of a book?

4   This clock shows the time in the afternoon. Write the time in two different ways.

5   A plane left Grenada at 15:40 and flew for 35 minutes. At what time did it arrive at its destination?

6   Look at the picture.

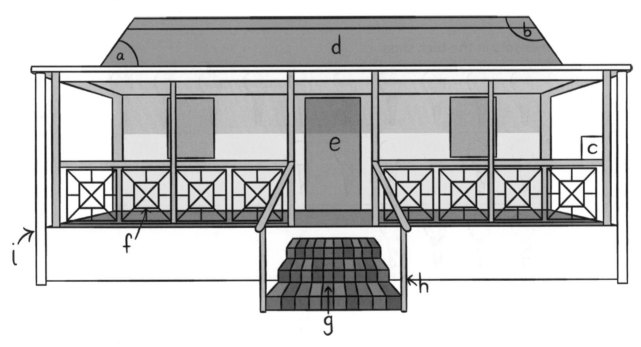

    a   Look at angles a to c. What type of angle is each?
    b   Name the plane shapes marked d to f.
    c   What solid objects are labelled g to i?

7 This thermometer shows the temperature in St Lucia.

a What is the temperature in St Lucia in degrees Celsius?

b At the same time, it is 70 °F in Florida. How many degrees cooler is that than the temperature in St Lucia?

8 A rectangular floor is 15 metres long and 8 metres wide.

a What is the perimeter of the floor?

b Calculate the area of the floor.

c How many square tiles with sides 50 cm long would you need to cover the floor?

9 This pictograph shows the number of ice-cream cones sold in the school's tuck shop.

Key: represents 2 cones

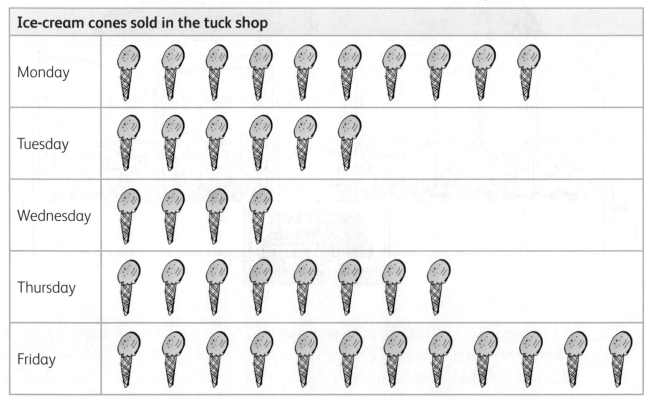

a   On which day did the tuck shop sell the most ice-creams?

b   Why do you think most ice-creams were sold on that day?

c   On which day did the tuck shop sell triple the amount sold on Wednesday?

d   What is the difference between the minimum and maximum number of ice-cream cones sold?

e   The cones are sold for $2.50. Calculate the total from sales on the 4th day of school.

f   On which day did the tuck shop sell $25 worth of ice-cream cones?

10   The table shows the marks (out of 40) students obtained in a test.

Mark out of 40	5	10	15	20	25	30	35	40
Number of students	0	1	2	5	8	9	5	2

a   Draw a bar graph to show the data.

Use your graph to answer parts b to e.

b   What was the total number of students in the class?

c   How many students scored less than 20 marks?

d   How many students scored 75% in the test?

e   What mark did 25% of the students obtain in the test?

11   A shoe shop kept a record of the shoe sizes sold one Saturday.

Shoe sizes purchased   4   5   6   7   3   7   6   5   7   7   9   6

a   Calculate the mean of the sizes.

b   Which shoe size was closest to the mean?

c   Which shoe size was furthest from the mean?

d   What is the modal size?

e   What is the median size?

12   A class teacher wants to find out what the most popular online games are that the students in her class are playing.

a   How would she collect this information?

b   What are some of the ways in which this information can be presented?

c   Work in pairs to conduct this survey in your class. Present your data in tables and graphs.

Teaching notes

Count in a higher number range

* Students should be able to use the patterns they already know to count beyond 99 999 and into millions.

Understand place value to millions

* For numbers above 99 999, the place value table is extended to the left to include a place for hundred thousands and then millions.

* Numbers can be written as a sum of the values in each place. This is called expanded notation. For example: 1 246 345 = 1 000 000 + 200 000 + 40 000 + 6 000 + 300 + 40 + 5.

* The position (place) of a digit affects its value. In the number 206 345, the digit 3 has a value of 300 because it is in the hundreds place. The 2 has a value of 200 000 because it is in the hundred thousands place.

Rounding numbers

* The rules for rounding off numbers apply to all places. Students need to understand how to use the digit to the right of the rounding place to decide how to round the number. If that digit is 5 or greater, you add 1 to the rounding place digit; if it is 4 or less you leave it as is. Then you fill in 0 in the places to the right of the rounding digit.

Large numbers in real life

* Students need to be able to read and understand large numbers in real-life situations. For example, the land area of different countries, population statistics, computer capacity, national debt and trade figures. The media is a good source of such numbers in use, but students will also come across these in their Social Studies and other textbooks.

* Many large numbers in real life are rounded off (estimated) to make them easier to read and make sense of them.

A

In one year, a shipping company transported 27 851 containers of bananas. The bananas had a combined mass of 286 902 tonnes. Can you say each of these numbers? Which number has 8 in the ten thousands place? What is the value of the 8 in the other number?

B

A large cruise ship can carry nearly eight thousand passengers and crew, and has a mass of over two hundred and twenty-five thousand tonnes. Write each of these numbers. Do you think these are exact numbers or estimates? Why?

1234

Where would you see a display like this one? What information does it give you? Work out what the next number will be after the display reaches 199 999.

Think, talk and write

A **Counting and place value** (pages 10–11)

1   Can you write these numbers in numerals?
    a   Twenty-three thousand five hundred and twelve
    b   Twenty-five thousand

2   What is the number?
    a   1 more than 999
    b   1 more than 9 999

3   Say each of these numbers aloud.
    a   65 000   b   32 345
    c   34 987   d   30 500
    e   14 502

B   **Rounding and estimating** (pages 12–13)

1   Nina says: 'There are about 400 children in my school.'
    a   Does this mean there are exactly 400 children in the school? Explain why or why not.
    b   Assuming Nina rounded the number of children correctly, could there be 350 children in the school? Why?
    c   Could there be 450 children in the school? Why?

2   Jayden has $25.00. He wants to buy three items costing $5.89, $3.90 and $15.65.
    a   How can he estimate to decide whether he has enough money?
    b   Does he have enough money?

C   **Large numbers in real life** (page 14)

1   Where do we use large numbers in real life? List at least five examples.

2   What is the largest number you have heard of, or used? Tell your group.

A Counting and place value

The **place value** table can be extended to the left to include higher and higher places. The place after ten thousands is called **hundred thousands**.

The number one hundred and twenty-five thousand is shown in the table.

We write this as 125 000.

Hundred thousands	Ten thousands	Thousands	Hundreds	Tens	Ones
1	2	5	0	0	0

In **expanded notation** this is:

125 000 = 100 000 + 20 000 + 5 000

The highest six-digit number we can write is 999 999.

We say nine hundred and ninety-nine thousand nine hundred and ninety-nine.

Maths ideas

In this unit you will:
* read and write numbers with up to 7 digits in numerals and words
* extend the place value table to include the hundred thousands and millions places
* state the place value and total value of any digit in numbers with up to 7 digits
* use expanded notation to write numbers
* arrange numbers in order of size.

1   Read the number names. Write each number in numerals.

   a   Fourteen thousand seven hundred and twenty-three

   b   Twenty-five thousand four hundred and seventy-four

   c   One hundred and sixty-three thousand four hundred and twelve

   d   Three hundred and seven thousand two hundred and eighty-nine

Key words

place value
hundred thousands
expanded notation
value
digit
millions

2   Write each number in numerals. Then say each number aloud.

   a   20 000 + 5 000 + 300 + 30 + 8   b   50 000 + 8 000 + 900 + 40 + 3

   c   300 000 + 80 000 + 3 000 + 800 + 90 + 3   d   200 000 + 90 000 + 4 000 + 700 + 20 + 5

3   What is the **value** of the **red** five in each number?

   a   23 5**0**8   b   5**4** 000   c   1**5** 876

   d   5**11** 500   e   234 0**9**5   f   49 0**5**0

This place value table shows what happens to the digits if we add 1 to 999 999.

Hundred thousands	Ten thousands	Thousands	Hundreds	Tens	Ones	
9^1	9^1	9^1	9^1	9^1	9	
					1	
1	0	0	0	0	0	0

$9 + 1 = 10$

You cannot have 10 in one column. Add 1 to the **digit** in the next column. Repeat this as you move across the table.

999 999 + 1 = 1 000 000   1 000 000 is one thousand thousands.

The mathematical name for one thousand thousands is one million.

We add the place for **millions** to the left of the hundred thousands column.

Millions	Hundred thousands	Ten thousands	Thousands	Hundreds	Tens	Ones
3	2	1	7	4	3	2

To read the number:
* read the millions first   three million
* read the thousands next   two hundred and seventeen thousand
* read the rest of the number   four hundred and thirty-two.

This number is three million two hundred and seventeen thousand four hundred and thirty-two.

4   Write the value of the **blue** digit in each number.

   a   3 243 809   b   4 530 987   c   3 198 087   d   1 209 999
   e   1 098 765   f   987 912   g   1 345 983   h   9 129 999

Problem-solving

5   Ms Dorleon wrote these numbers on the board.

1 250 000   1 899 245
     3 525 000   3 255 000
1 499 300   1 465 250
        1 899 765   1 876 125
1 350 425   3 552 000

A   My number has 5 hundred thousands and 5 ten thousands.

B   Mine has more than 2 tens and 6 ten thousands.

C   I've got a number that is greater than three million five hundred thousand with a 2 in the thousands place.

D   My number is less than 1 500 000 and it has 3 hundreds.

   a   Four students each chose
        a number from the board.
        Read what they say and work out which number each student chose.

   b   Ms Newton drew this diagram to sort the numbers.

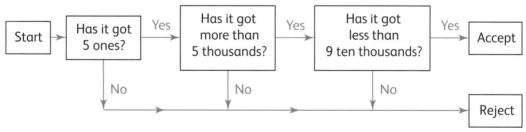

        Which of the numbers on the board will be accepted by the machine?

What did you learn?

Write in numerals the number that is:

1   1 million less than 2 500 000   2   200 000 more than 3 million

3   55 000 more than 2 411 000   4   500 000 less than 2 600 000

B Rounding and estimating

Maths ideas

In this unit you will:
* revise the rules for rounding numbers
* round numbers to the nearest ten, hundred or thousand
* use rounded numbers to estimate answers.

Explain

Do you remember how to use **place value** to **round** numbers to a given place?
* Find the **digit** in the rounding place.
* Look at the digit to the right of this place.
* If the digit to the right is 0, 1, 2, 3 or 4, leave the digit in the rounding place as it is.
* If the digit to the right is 5, 6, 7, 8 or 9, add 1 to the digit in the rounding place.
* Change all the digits to the right of the rounding place to 0.

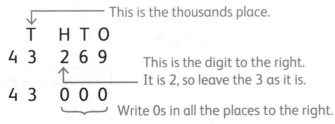

Round 43 269 to the **nearest** thousand.

This is the thousands place.

T H T O
4 3 2 6 9   This is the digit to the right.
               It is 2, so leave the 3 as it is.
4 3 0 0 0
               Write 0s in all the places to the right.

Round 18 763 to the nearest hundred.

This is the hundreds place.

1 8 7 6 3
               This digit is more than 5,
               so we change the 7 to 8.

1 8 8 0 0   Write 0s in all the places to the right.

Key words

place value	nearest
round	estimate
digit	

Think and write

How would you round these numbers to make them more user-friendly?

1 There are 12 309 students in our school district.

2 Last year 1 340 876 tourists visited the Caribbean during winter.

3 The population of an island is 387 497.

1   Round these numbers to the nearest hundred and to the nearest thousand.

   a   23 987   b   13 876

   c   232 045   d   19 167

   e   897 302   f   276 540

   g   99 087   h   129 000

Problem-solving

2   The population of each country below has been rounded to the nearest thousand.

Country	St Lucia	Barbados	Trinidad and Tobago	Jamaica	Puerto Rico
Population	172 000	283 000	1 357 000	2 729 000	3 508 000

   a   What is the minimum and maximum number of people that could be in each country?

   b   Which country has the closest to 3 million people?

Explain

Estimating is a very useful strategy to help you calculate quickly and to help you decide whether your answer is reasonable or not. You can use rounding to **estimate** an approximate answer.

Estimate 468×62:
500×60 Round each number to the first digit.
$5 \times 6 = 30$ Use the facts you already know.
so $500 \times 60 = 30\,000$

$468 \times 62 \approx 30\,000$ $\approx$ means 'approximately equal to'
When you give an estimated answer, you use the $\approx$ symbol.

Challenge

3 Use the figures from the population table on page 12, and the rules you know for rounding numbers.

  a Round the populations of Trinidad, Jamaica and Puerto Rico to the nearest million.

  b Which country has a population of 300 000, if estimated to the nearest hundred thousand?

4 Estimate each of the following.

  a $39 + 42$ b $499 - 67$ c $32\,876 - 12\,909$ d $29 + 187$

  e $148 - 9 + 24$ f 58×22 g 32×12 h $32\,876 + 13\,087$

Problem-solving

5 The number of cruise passengers passing through a terminal during a week is given in the table.

Day	Monday	Tuesday	Wednesday	Thursday	Friday
Number of passengers	13 689	25 908	18 579	18 798	21 809

  a   Round the passenger numbers to the nearest thousand and then estimate the total number over five days.

  b   Use the figures rounded to the nearest thousand and estimate:

     i   how many more passengers passed through on Friday than on Thursday

     ii   the number of passengers there were on the two busiest days

     iii   the difference in passenger numbers between the busiest and least busy days.

6 Crowd attendance at a cricket match over a three-day period was 4 146, 5 964 and 7 193. Estimate the total attendance.

7 Approximately how many sets of 180 counters could you make from a box containing 24 275 counters?

What did you learn?
Round each price to the nearest thousand and estimate the total amount.

$8 632 $2 745 $9 034 $12 634 $10 999

C Large numbers in real life

Explain

You will often hear or read large numbers in your daily life. For example, you might hear **data** about **population** numbers or **tourism** figures. You may also hear very large amounts of money in the news. Governments report on **finances** when they collect taxes or invest in large projects.

Key words

data	tourism
population	finances

1   The cards show the population of some European countries in 2017.

Macedonia 2 108 434	Sweden 9 631 261	Finland 5 443 497	Norway 5 091 924
Austria 8 526 429	Denmark 5 640 184	Switzerland 8 157 896	Bulgaria 7 167 998

   a   Write the numbers in descending order.

   b   Which countries had fewer than 6 million people?

   c   Which countries had more than 8 million people?

   d   How many countries had between 5 and 6 million people?

2   A tourism website reported that 5 940 000 tourists visited the Eastern Caribbean in 2016 and 6 151 000 tourists visited the area in 2017.

   a   Write a statement using $<$ or $>$ to compare the numbers.

   b   In 2017 approximately 2 million of the tourists visited at least three islands. Approximately how many did not?

   c   In the first six months of the year, there were 3 828 415 tourists. If the same number visited in the second six months, would there be more than 7 million tourists? How did you work this out?

3   An electricity meter uses small dials like these to measure electricity usage in kilowatts. Each dial represents one place in the number.

 kW

   a   What is the reading on the dial that shows hundreds?

   b   Why do the dials only go up to 9?

   c   Use the dials to write down how many kilowatts have been used.

   d   The next month, the meter records 10 000 more kilowatts. What is the reading for that month?

What did you learn?

Why is it important to think carefully when you hear large numbers being used in the news or other places in everyday life?

Topic 2 Review

Key ideas and concepts

Read the statements. Fill in the missing words to summarise what you learnt in this topic.

1   A number in the _____ has seven digits.
2   Each digit in a number has its own _____ value.
3   Writing a number as a sum of place values is called _____.
4   Descending order means the numbers _____ in value.
5   To round a number to the nearest thousand, you look at the digit in the _____ place.
6   When you round numbers in a calculation you get an _____ answer.
7   _____ data is one example of large numbers used in everyday life.

Think, talk, write ...

1   Explain in your own words how to:
   a   say the number 3 876 016
   b   compare the numbers 3 456 089 and 3 465 908
   c   work out whether a number is in the millions or not.

2   Give three examples of where you might use or hear large rounded numbers in everyday life.

Quick check

1   Write the number you would get if you:
   a   added 5 hundreds to 14 499   b   added a million to 324 000.
2   Write each of these numbers in expanded notation.
   a   23 456   b   4 000 000   c   876 000   d   2 345 009
3   What is the value of the **red** digit in each of these numbers?
   a   2**3** 457   b   8 **7**65 432   c   **9** 876 309   d   2 309 **7**65
4   Use all the digits on the card without repeating any to make:
   a   the greatest possible number
   b   the smallest possible number
   c   the largest number with 9 in the thousands place
   d   the smallest number with 3 in the millions place.

   9   6   8
   5   7   2   4

5   The areas of the six biggest countries in South America are given in the table.

Country	Argentina	Bolivia	Brazil	Colombia	Peru	Venezuela
Area (km²)	2 766 890	1 098 580	8 514 877	1 138 910	1 285 220	916 445

   a   Which country is the biggest?
   b   Which country has an area of less than a million square kilometres?
   c   Which country has an area of more than 2 million square kilometres, but less than 3 million square kilometres?
   d   Write the areas in order from smallest to greatest.
   e   Round each area to the nearest 1 000.
   f   Use your rounded figures to estimate the combined area of the six countries.

Teaching notes

Mental strategies

* Continue to start each lesson with a short mental activity whenever possible to give students a chance to use number facts and consolidate strategies. Students only develop fluency and confidence if you provide regular and ongoing practice.

* Encourage mathematical thinking and reasoning. This involves making time to talk about how students worked out their answers mentally. Let students verbalise and explain the steps they took. Others will learn from these explanations as they become exposed to different options and methods. This allows them to consider multiple strategies and choose the one they find easiest and most efficient.

Pen-and-paper calculations in a higher number range

* Students already know how to add and subtract larger numbers and they have learnt how to do long multiplication and long division.

* Encourage students to use grid methods and expanded notation if they find these easier. Multiple strategies help students understand the mathematics. They will move to the most efficient methods of writing calculations when they are ready.

Problem-solving strategies

* Students need to realise that some problems have more than one step. They may need to calculate one value first and then use it in the next step.

* When you solve problems with the class, model the steps and talk students through your method of working. This makes problem-solving strategies visible to students who have not yet fully understood them.

A

What types of numbers can you see in the photograph? Which sale offer means the same as half price? How would you mentally work out the sale price of an item with 50% off? What does 25% off mean?

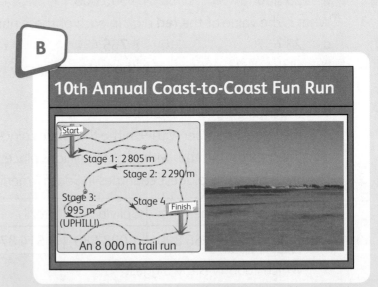

B

10th Annual Coast-to-Coast Fun Run

Start
Stage 1: 2805 m
Stage 2: 2290 m
Stage 3: 995 m (UPHILL!)
Stage 4
Finish
An 8 000 m trail run

Read the information on the map. How long is the whole run? How far will you have run if you have completed the first three stages? The length of Stage 4 is not given. How would you work it out?

C

A trip on a glass-bottom boat costs $15 per adult and $12 per child. How would you work out the cost for a group of 7 adults and 6 children? A group of school children go on the boat. The tickets cost $168. How would you work out how many children there were?

D

Mr Johnson grew one of these vegetables. The one he grew was not red or green and it was not round. Which one did he grow? What strategy did you use to work out the answer?

Think, talk and write

A Mental strategies (pages 18–19)

* Marcus has to work out 5×28 in his head. This is how he thinks: What strategy did Marcus use to calculate the answer?

> 5 is half of 10.
> 10×28 is 280.
> Half of 280 is 140.

* Michelle has to calculate 12×25. This is how she thinks: Why do you think she halved the 12 and doubled the 25?

> 12×25 is hard.
> 6×50 is easier.
> 6×5 is 30.
> 30×10 is 300.

* How would you work out 6×18 in your head? Write a thought bubble to show your thinking.

B Addition and subtraction (pages 20–22)

1 Use the figures on the map on page 16 to work out:
   a the difference in distance between Stage 1 and Stage 2
   b the total distance from the end of Stage 1 to the start of Stage 4
   c how much further you have to run in Stage 2 than in Stage 3.

2 Joe trains for the race by running the Stage 3 route four times each day for a week. Micah trains by running the Stage 1 route twice a day for a week.
   a Who runs further?
   b How much further do they run?

C Multiplication and division (pages 23–25)

Discuss these questions in groups.

1 How do you know when a problem involves multiplication?

2 How do you know when to divide to solve a problem?

3 Paul says $174 \div 23$ equals 6 remainder 35. How do you know he is wrong without checking the calculation?

4 Kezia works out 124×20 and gets an answer of 248. What has she done wrong?

D Problem-solving strategies (pages 26–28)

Discuss how you would work to solve this problem. Last weekend, 4 202 people attended a concert. This is 11 times more than the number of musicians who performed at the concert. How many musicians were there?

A Mental strategies

A **mental** calculation is one that you work out in your head, by thinking, rather than using pen-and-paper methods. Sometimes you might need to jot down steps on scrap paper. This is not the same as doing long calculations: **jottings** are just reminders, not workings.

Read through these **strategies** and examples carefully.

Maths ideas

In this unit you will:
* revise some of the strategies you have learnt to add, subtract, divide and multiply mentally
* develop some mental strategies for finding percentages or fractions of amounts.

Key words

mental	strategies
jottings	percentages

Multiply by powers of 10 (10, 100, 1 000, and so on)

Put zeros at the end of the number to show how many places the digits have moved:

23 × 10 = 230 and 23 × 1 000 = 23 000

Multiply by 2

Double each digit in the number. This is the same as adding the number to itself:

123 × 2 = 123 + 123 = 246

Multiply by 5

Multiply by 10 (put one zero on the number) and then halve the result:

$5 \times 23 = \frac{1}{2}$ of 10 × 23

10 × 23 = 230

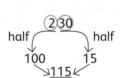

Divide by 2

Halve the values in each place or group of places:

518 ÷ 2

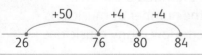

Multiply by single-digit numbers

Use doubling and halving when you can:

5 × 46 = 10 × 23 = 230

7 × 18 = 7 × 9 × 2 = 63 × 2 = 126

Add two-digit numbers

Use place value and add in parts:

23 + 34 = 20 + 30 + 3 + 4 = 57

25 + 36 = 20 + 30 + 5 + 6 = 50 + 11 = 61

Subtract two-digit numbers

Subtract in parts if you can:

73 − 21 ⟶ 73 − 20 ⟶ 53 − 1 = 52

Use chunking to count up or back (with or without a number line):

84 − 26

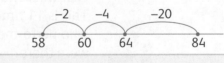

Percentages or fractions of amounts

Remember: **Percentages** are a type of fraction and the word 'of' tells you to multiply.

50% of a number means the same as $\frac{1}{2}$ of the number, and you can divide by 2.

25% of a number means the same as $\frac{1}{4}$ of the number, and you can divide by 2, and then by 2 again.

10% of a number means the same as $\frac{1}{10}$ of the number, and you can divide by 10.

1 Use the mental strategies you find easiest to do these calculations.

a	43 + 20	b	64 + 20	c	73 + 30	d	29 + 60
e	43 + 21	f	63 + 42	g	71 + 19	h	34 + 43
i	36 + 15	j	47 + 26	k	37 + 35	l	43 + 58

2 Calculate mentally. Write the answers only.

a	43 − 10	b	65 − 30	c	87 − 20	d	49 − 30
e	49 − 21	f	67 − 23	g	73 − 23	h	98 − 43
i	83 − 25	j	63 − 35	k	72 − 46	l	54 − 38

3 Calculate mentally. Write the answers only.

a	23×10	b	19×100	c	45×10	d	123×100
e	32×5	f	12×18	g	25×8	h	32×5
i	14×25	j	20×23	k	14×15	l	84×50

4 Calculate.

a	50% of $84	b	10% of 120	c	10% of $450	d	25% of 420

Investigate

5 Shirleen says that if you want to multiply by 25 mentally, you can times by 100 and then halve the answer and halve it again. Test Shirleen's method on a few examples. Explain why it works.

What did you learn?

1 Calculate mentally. Write the answers only.

a	50×38	b	17×50	c	40×16
d	56×25	e	76×25	f	25×32

2 What is:

a	47 less than 73?	b	56 less than 93?	c	25% of 84?
d	43 more than 32?	e	48 more than 19?	f	10% of 920?

B Addition and subtraction

In this unit you will:
* use written methods to add and subtract numbers
* understand how to regroup numbers in any place
* use rounded or compatible numbers to find approximate answers (estimates)
* use estimates to decide whether an answer is reasonable or not.

Explain

You already know how to **add** and **subtract** numbers using **expanded notation** and column methods.

You also know that it is important to **estimate** before you add or subtract so that you can decide whether your answer is reasonable or not. You can round numbers to a suitable place value to estimate. You can also estimate **sums** and **differences** using compatible numbers.

A **compatible number** is one that is fairly close to the value you are working with, but it is easier to add or subtract.

Compatible numbers can give you an estimate that is closer to the real value than rounded numbers.

Key words

add
subtract
expanded notation
estimate
sum
difference
compatible number
regroup

Example 1

Use compatible numbers to estimate the answer to each calculation.

a $455 + 223 \approx 670$
    $450 + 220 = 670$
    450 and 220 are compatible, so you can add them mentally.

Rounding the values to the leading place would give you $500 + 200 = 700$. The compatible numbers give you 670, which is closer to the real value of 678.

b $845 - 637 \approx 200$
    $850 - 650 = 200$
    850 and 650 are compatible, so you can subtract them mentally.

Read through the next example to remind you how to add and subtract larger numbers in columns.

Example 2

Regroup by carrying

$$\begin{array}{r} {}^{11}\ \ {}^{11} \\ 12\ 876 \\ +\ 9\ 429 \\ \hline 22\ 305 \end{array}$$

Regroup and carry

$$\begin{array}{r} {}^{1}1\ \ {}^{7}\cancel{8}^{1}5 \\ 2\cancel{4}\ 6\cancel{8}5 \\ -19\ 437 \\ \hline 5\ 248 \end{array}$$

---

1   Estimate and then add.
    a $142 + 117 + 131$
    b $289 + 2\,114 + 1\,309$
    c $12\,345 + 4\,568 + 99$
    d $1\,456 + 7\,654 + 8\,123$
    e $765 + 1\,234 + 678 + 99$
    f $43\,568 + 12\,897$

2   Estimate and then subtract.
   a   8 798 – 1 307
   b   8 032 – 4 156
   c   12 345 – 8 765
   d   321 098 – 158 987
   e   88 950 – 43 999
   f   121 625 – 112 887

3   Haiti and the Dominican Republic are two countries on the same island. The area of Haiti is 27 560 square kilometres and the area of the Dominican Republic is 48 320 square kilometres.
   a   What is the total area of the island?
   b   What is the difference between the areas of Haiti and the Dominican Republic?

4   The lengths of some of the world's longest rivers are given below.

River	Amazon	Nile	Yangtze	Mississippi	Yellow
Length (km)	6 992	6 853	6 418	6 275	5 464

   a   What is difference in length between the longest and the shortest river?
   b   The Yangtze River and the Yellow River are both in China. What is their combined length?
   c   How far would you travel if you sailed all the way up and down the Nile?
   d   A researcher wants to travel the length of all five rivers. What is this distance?
   e   Make up two addition and subtraction problems using data from the table. Exchange your problems with another student. Check each other's answers once you have solved the problems.

Think and talk

5   Sort the words and phrases in the box below into words that tell you to subtract and words that tell you to add.

   | minus   deduct   combine   total   sum   find the difference   how many less   how many more   take away   altogether   how many in all   how much |

   Can you think of any other words that tell you whether you should add or subtract to solve a problem?

Explain

Sometimes you need to do more than one operation to solve a problem.

Example

Maria buys old furniture and sells it to make a profit. The difference between what she pays for it and what she sells it for is her profit.

Maria paid $420.00 for two desks. She sold one for $348.00 and the other one for $364.00. What was her profit on the two desks?

Think: Her profit is what she sold them for minus what she paid.

Sold for: Minus what she paid:
   $348.00 $712.00
 + $364.00   – $420.00
   $712.00 $292.00   Her profit was $292.00.

6   Maria bought a table for $329.00 and a chair for $89.00. She later sold them as a set for $700.00. What was her profit on that sale?

7   The marks out of 100 for four students in three exams are given below.

Student	English	Mathematics	Science
Jonelle	76	82	72
Keshawn	75	81	79
Kezia	81	79	77
Leroy	74	78	76

   a   What is the total score out of 300 for each student?

   b   What is the difference between the highest and the lowest total score?

8   The land area of some large islands is given here in square kilometres.

MADAGASCAR 578 041

GREENLAND 2 130 800

UNITED KINGDOM 216 777

BAFFIN 507 451

NEW GUINEA 800 311

CUBA 110 860

   a   The Caribbean Sea has an area of 2 754 000 square kilometres. How much larger is this than the area of each island?

   b   Which is larger: Baffin Island or Madagascar, and by how much?

   c   Which two islands have a combined area of 1 million square kilometres when it is rounded to the nearest million?

What did you learn?

1   Estimate and then calculate.
   a   190 991 + 12 234 + 14 568   b   3 132 819 + 2 343 214   c   2 112 345 + 6 123 145
   d   113 285 − 29 873   e   3 229 876 − 1 314 388   f   8 234 000 − 1 939 453

2   The solutions to five different problems are given below. For each one, write a word problem that fits the number sentences.
   a   23 456 − 12 987 = 10 469   b   463 + 587 = 1 050   c   1 050 − 209 = 841
   d   345 + 245 + 650 = 1 240   e   1 845 − 1 240 = 605

C Multiplication and division

Explain

You learnt how to **multiply** and **divide** larger numbers using pen-and-paper methods last year.

Remember to **estimate** before you calculate. Use the method that suits the numbers you are working with.

Read through the examples to help you remember how to do this.

Maths ideas

In this unit you will:
* use pen-and-paper methods to multiply and divide using larger numbers
* solve problems that involve multiplication and/or division.

Key words

multiply	dividend
divide	divisor
estimate	remainder
product	decimal
quotient	

Example 1

Find the **product** of 497×43.

Long method

Estimate: $500 \times 40 = 20\,000$

$$
\begin{array}{r}
{}^{3}2\,{}^{2}4\,{}^{2}9\;7 \\
\times\qquad 4\;\;3 \\
\hline
{}^{1}1\;{}^{1}4\;9\;1 \\
{}^{1}1\;9\;8\;8\;0 \\
\hline
2\;1\;3\;7\;1
\end{array}
$$

$\times$ 43 is the same as ($\times$ 40) + ($\times$ 3)

This is 3×497.

If you multiply by 40, you will get one zero at the end of the answer. Write this in the ones place. Then work out 4×497.

Grid method

$\times$	400	90	7
40	16 000	3 600	280
3	1 200	270	21

$\quad$ 19 880
+ 1 491 = 21 371

Example 2

How many times does 4 go into 784?

Estimate: $800 \div 4 = 200$

Short division:

$$
\begin{array}{r}
1\;9\;6 \\
4\overline{)7\,{}^{3}8\,{}^{2}4}
\end{array}
$$

Example 3

What is $867 \div 21$? $\qquad$ Estimate: $860 \div 20 = 43$

Long division:

$$
\begin{array}{r}
41\text{r}6 \\
21\overline{)867} \\
-84\downarrow \\
\hline
27 \\
-21 \\
\hline
6
\end{array}
$$

$21 \times 4 = 84$

$21 \times 1 = 21$

The answer to a division is called the **quotient**. The number being divided is called the **dividend** and the number you are dividing the dividend into is called the **divisor**.

When one number does not divide exactly into another, you are left with a **remainder**.

1   Estimate and then calculate using pen-and-paper methods. Check your answers using
    a calculator. If you make a mistake, try to find it.

    a 3 006 × 8   b 2 406 × 9   c 3 008 × 7
    d 87 × 98   e 312 × 45   f 423 × 87
    g 145 × 208   h 876 × 39   i 412 × 602
    j 342 × 456   k 298 × 342   l 612 × 43
    m 4 123 × 45   n 1 467 × 234   o 12 098 × 312

2   Estimate and then calculate using pen-and-paper methods. Check your answers using
    a calculator. If you make a mistake, try to find it.

    a 800 ÷ 7   b 345 ÷ 6   c 872 ÷ 8
    d 786 ÷ 24   e 832 ÷ 25   f 347 ÷ 23
    g 1 345 ÷ 21   h 3 214 ÷ 13   i 5 330 ÷ 65
    j 12 765 ÷ 27   k 14 235 ÷ 70   l 10 819 ÷ 51

Explain

You can divide money amounts with **decimal** parts in the same way as you divide whole numbers.
Remember to line up the decimal points in both the answer and the working part of the division.

3   The total cost for different activities is given for a number of people. Work out the cost per person.
    a Helicopter ride for 4 adults: $1 245.80
    b Hot air balloon trip for 5 people: $1 241.75
    c Parasailing and snorkelling for 3 people: $218.97
    d Deep-sea fishing for six people: $2 036.40
    e Waterpark and lunch for four people: $254.60

Problem-solving

For each problem, estimate before you solve and check that
your answer is reasonable.

4   Dina has a roll of red ribbon that is 3 600 cm long. She
    uses it all to wrap 15 presents by cutting equal lengths of
    ribbon for each gift. How long are the pieces she cuts?

5   Steve is tying big bows. He has a piece of blue cord
    381 cm long. He cuts 23 pieces of equal length and has
    13 cm left over. How long is each piece that he cut?

6   A school textbook has 128 pages. How many pages are there in 502 of these books?

7   A factory makes 624 rum cakes each week. They close for two weeks over Christmas.
    How many cakes do they make per year?

8   A farmer has 1 276 pineapples. She wants to pack them into 58 crates. Each crate
    should have the same number of pineapples. How many pineapples should be
    packed into a crate?

9 460 students took part in a parade. They were placed in groups of 24. How many groups of 24 could the organiser make? What do you think could be done with the remaining students?

10 Jasmine gets $10.00 per week for doing chores. She spends $8.25 each week and saves the rest. How long will it take her to save $25.00?

11 A car repair shop ordered 23 motors costing $257.00 each. What will the total cost be?

12 Hyacinth's car uses a litre of gas per 12 kilometres on average. How many litres will it use if she travels 850 kilometres?

13 There are 36 packets of chips in a box. Each packet costs 80 cents.
   a What does a box cost?
   b A school tuck shop sells 1 100 packets of chips per week. How many boxes should they order each week?
   c What would the cost of the weekly order be?
   d The tuck shop sells the chips for $1.00 per packet. How much profit do they make per week?

14 The grandstands at a sporting venue have 48 rows with 52 seats per row. How many seats are there?

15 A ship used 8 760 litres of fuel for 12 return journeys of the same distance. How many litres is this per single journey?

16 Ticket prices for a waterpark are displayed at the entrance.
   a What would the tickets cost for a group of 9 adults, 25 children and 32 students?
   b The total cost for a group of students aged 12 to 16 was $896.00. How many students were in the group?

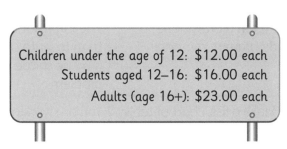

Children under the age of 12: $12.00 each
Students aged 12–16: $16.00 each
Adults (age 16+): $23.00 each

What did you learn?

1 Estimate and then calculate. Show all your working.
   a 325 × 6
   b 754 × 38
   c 107 × 39
   d 98 ÷ 17
   e 130 718 ÷ 6
   f 34 213 ÷ 34

2 Tony has 2 462 bags of mangoes. He packs them into crates that can hold 32 bags. How many crates can he fill? How many bags of mangoes will be left over?

Challenge

3 A rectangle is 14 times as long as it is wide. If it is 23 cm wide, calculate its perimeter and its area.

D Problem-solving strategies

Explain

To solve problems, you first need to read and understand the problem, and then you have to choose a suitable **strategy** to solve it. Next, you work out the **solution** and, finally, you check that it seems **reasonable** and correct.

There is often more than one way to solve a problem. Many problems can be solved just by thinking and working systematically, but others may need you to write a **number sentence** (**equation**) to represent the problem and then work out the solution mathematically.

Read the information in the table and think about the types of problems that could be solved using each strategy.

Maths ideas

In this unit you will:
* revise problem-solving strategies
* choose and use the most appropriate strategy to solve a range of different problems.

Key words

strategy	number sentence
solution	equation
reasonable	inverse

Strategy	When is it useful?	Tips
Trial and error, also called 'guess, check and refine'	You are not sure of the answer, but you think you have some options that you could try. Example 1 on page 27 uses this strategy.	Do not make wild guesses. Use the information you have to guess sensibly. Check if the answer is correct. If not, use what you have learnt from your guess to get closer to the correct answer.
Sketch a drawing, a diagram or a model	The information is easier to sketch than to write out. These problems may be shape or measurement problems.	Label your sketch properly with all the necessary information.
Make a list, chart or table	You are working with different combinations, probabilities or sets of data. Example 2 on page 27 uses this strategy.	Choose the option that best matches your information.
Eliminate possibilities	There are only a few possible correct answers. Eliminate those that do not work.	Use tables or grids. Cross off answers as you eliminate them.
Use **inverse** operations	The problem calculation uses one of the four operations. However, some information is missing.	Use the information you already have to set up the inverse operations.
Write a number sentence (equation)	The problem gives numerical information and some clues about what operations you would need to solve it.	Always set out your working clearly, say what any letters represent and show any working out that you do.

Example 1

Dennis's age is $\frac{1}{5}$ of his dad's age. The sum of their ages is 54. How old is Dennis's dad?

Try 40	Guess	Try 45	Guess
$\frac{1}{5}$ of 40 = 8	Check	$\frac{1}{5}$ of 45 = 9	Check
40 + 8 = 48	Refine: this is too low, so try again	45 + 9 = 54	That's the sum of their ages, so Dennis's dad is 45.

Example 2

How many whole numbers less than 100 contain the digit 5?

Five can be in the ones place: 5, 15, 25, 35, 45, 55, 65, 75, 85, 95

Five can be in the tens place: 50, 51, 52, 53, 54, ~~55~~, 56, 57, 58, 59

Do not count 55, as it is in the first list.

10 + 9 = 19 $\longrightarrow$ There are 19 whole numbers less than 100 with the digit 5.

Problem-solving

Work in pairs or small groups. Decide which strategy is most useful for solving each problem. Then work on your own to solve the problems.

1   This is a combination lock padlock. Each round section has all the digits from 0–9 on it. You choose three digits as your passcode, and only those three in the correct order will unlock the padlock.

   a   How many different passcodes do you think you can you make if you do not repeat any digits? Discuss your ideas with your group.

   b   Jania set a code on a padlock like the one shown here. She wrote these clues down so that she would remember what the passcode is. Can you work out what it is?

> It has three different odd digits.
>
> If it was a 2-digit decimal, it would round to 5.8.
>
> It is divisible by 3, but not by 9.

2   Angela bought a present and a card for her friend. The present cost $16.00 more than the card and she paid $25.00 in total. What was the cost of the card?

3   Diagram A contains 5 squares. How many squares are there in diagrams B and C?

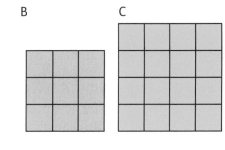

A

4 $\rightarrow$ 1 × 1 squares

1 $\rightarrow$ 2 × 2 square

B   C

4 The numerals 1 to 7 are written on seven different cards.
   a How many pairs of cards add up to 8?
   b How many ways are there to group three cards and get a total of 10?

5 A large boat-building company makes 15 dinghies per day, five days per week, for 48 weeks of the year. How many dinghies do they make in a year?

6 If everyone arriving at a party shakes hands with every other person there, how many handshakes will there be altogether if there are:
   a 2 people?   b 5 people?   c 10 people?
   d Can you find a method to work out how many handshakes there will be for any number of people (*n* people)?

7 Write an equation to represent each problem and then solve it to work out the unknown value.
   a 15 more than *x* is 20   b 10 less than *a* is 30   c 15 minus *b* is 10
   d half of *x* is 22   e twice *y* is 30   f the square of *x* is 16

8 Grandma Smith tells her grandson that she is 60 years old and that she married Grandpa Smith 34 years ago. She got a Master's Degree 3 years after she got married. How old was she when she got her Master's Degree?

9 T-shirts on a stall are $8.60, $9.20, $10.80 or $12.40, depending on the size. A customer bought six T-shirts. Which of these amounts could be the total cost?
   a $63.30   b $74.50   c $66.60

Challenge

10 Mike, Pete, James and Charles live in a four-storey apartment block. The floors are numbered from 1st to 4th.
   * Their surnames are Brown, White, Grey and Johnson, but not necessarily in that order.
   * Mike lives two floors below James.
   * Pete lives on the top floor.
   * Mr Brown lives on the third floor.
   * Mr White lives on the floor above Mr Grey.
   What is the name and surname of the person on the second floor?

What did you learn?

1 Mr Nixon has 2 lengths of rope. One is 1.895 m long, the other is 1.904 m long.
   a Which piece is longer and by how much?
   b If he needs 4.5 metres of rope in all, how much more will he need to buy?

2 a How many ways can you make $1 using any combination of 10, 20, 25 and 50 cent coins?
   b Why did you choose the strategies you used to solve this problem?

Topic 3 Review

Key ideas and concepts

Copy the unit headings from this topic into your book. Write short notes to summarise the main things covered in each unit.

A   Mental strategies

B   Addition and subtraction

C   Multiplication and division

D   Problem-solving strategies

Think, talk, write …

Discuss the questions in groups and then write your own answers in your maths journal.

1   Did you learn any new methods of working in this topic?

2   Why is mental mathematics useful?

3   Besides at school, can you think of three examples where you might do a mental calculation?

4   Which mental strategies do you find easiest to use? Why?

5   Did you struggle with any of the problem-solving strategies? Why?

Quick check

1   Look at the prices of different tours.

Work out the price of these tours using mental strategies.

a   Fishing Fun and Creepy Caves

b   Island Hopping and Creepy Caves

c   Volcano Visit for 10 people

d   Creepy Caves for 5 people

e   A combined Fishing Fun and Island Hopping tour with 10% off the total.

Tour	Price per person
Fishing Fun	$99
Volcano Visit	$180
Creepy Caves	$250
Island Hopping	$420

2   Calculate.

a $12\,345 + 9\,987$

b $12\,098 - 3\,945$

c $23\,412 \times 9$

d $4\,123 \times 45$

e $2\,314 \div 11$

f $43\,126 \div 21$

3   How many 55-seater buses are needed to transport 857 people?

4   Write a problem to match this number sentence.
$\$27.50 + (\$250 \times \$3.00) = \$777.50$

5   In how many different orders can you write the words 'good', 'better' and 'best'?

6   A carpenter plans to make 22 kitchen stools. Each stool can either have 3 or 4 legs. He has 81 legs in stock and he wants to use them all. How many of each type of stool can he make?

7   A number less 9 is equal to the product of 5 and 8. What is the number?

Topic 4
Shape and space (1)

Teaching notes

Angles and lines

* Students should know the difference between a point, a line segment, a line and an angle. They should also know the terms parallel and perpendicular.

* This year, students will learn how to read and use a protractor to measure angle sizes in degrees. You need to demonstrate how to use a protractor and help students who find this difficult. There are many online resources and videos that you could use.

2-D shapes

* Remind students that plane or 2-D shapes have two dimensions only: length and breadth. Avoid the term flat shapes, as this can be misleading. A piece of paper or a sticker is a flat shape, but mathematically it has a third dimension (depth), so it is not a plane shape.

* Revise the concept of a polygon and the names of common polygons. Also revise circles and the names of circle parts.

Quadrilaterals

* All four-sided shapes are quadrilaterals. Students already know that there are different types (they should be very familiar with squares and rectangles).

* This year students will learn the names of other types of quadrilaterals and use the special properties of each shape to identify and classify it.

Classifying triangles

* Students already know that triangles have different angle sizes. Revise the terms acute and obtuse if you need to.

* Students will learn in this topic how to use the lengths of sides to name and classify triangles. They need to know the terms equilateral, scalene and isosceles.

A

Look at the picture carefully. Find as many right angles as you can. Show them to your partner. Can you find angles that are smaller or bigger than right angles? Where are they? Find examples of parallel and perpendicular lines in the picture.

B

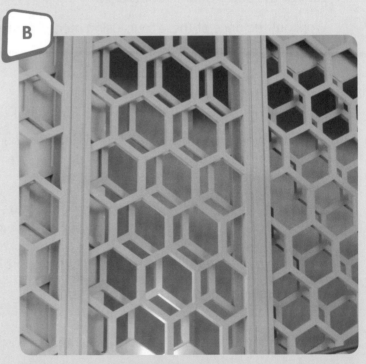

This is a design on the wall of a large hotel. Look at the central panel. What shapes were used to make the design? Would this design work with octagons? Why or why not? What other shapes have been made by overlapping the main shapes and by cutting them off?

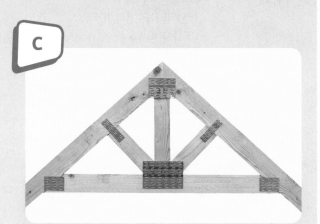

C

How many triangles can you find in this picture? What is the same in all the triangles? What differences can you find between them?

D

This glass pyramid with four faces is outside a famous museum called the Louvre in Paris, France. What shape is the base of the pyramid? What shape are the four faces? What shapes have been used to make the glass panels?

Think, talk and write

A Angles and lines *(pages 32–35)*

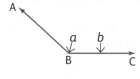

1   Explain what this diagram shows.

2   Name the parts marked *a* and *b*.

3   Does the length of part *b* affect the size of the angle? Explain your answer.

4   Draw diagrams to show a point, a line and a line segment.

B Shapes and their properties *(pages 36–39)*

1   What do the following mean when applied to 2-D shapes?
   a   tri-   b   quad-   c   hex-   d   octa-

2   'Quad' means four and 'lateral' means side. So what does 'quadrilateral' mean?
   a   Find five examples of quadrilaterals around you. List them and draw a sketch of each.
   b   How are the quadrilaterals that you found similar?
   c   How do they differ?

C Classifying triangles *(pages 40–41)*

Look at the diagram and answer the questions.

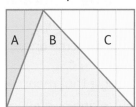

1   What do triangle A and C have in common?

2   Which triangle has two equal sides?

3   What types of angles does triangle B have?

4   Sam says triangle A and C are the same type. Nikita says they are two different types of triangle. What do you think?

D Shapes around us *(page 42)*

How do artists and designers use angles, lines and shapes in their work? Discuss this in groups and try to find some local examples to show what you mean.

A Angles and lines

Do you remember what you learnt last year about **points**, **lines** and **angles**?

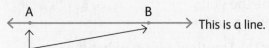

This is a line.

A and B are points on the line.

AB is a **line segment** because it has two end points.

When lines meet or cross at a point they form an angle. The point where the lines meet is called the **vertex**. The sides of the angle are called its **arms**.

This angle is named AB̂C, or CB̂A. Point B is the vertex, so it must always be in the middle when you name the angle.

The vertex can also be used on its own to refer to the angle: B̂.

Angles are classified by size.

An angle that measures exactly 90° is a **right angle**. A right angle looks like the corner of a square. IĴK is a right angle.

An angle that measures between 0° and 90° is called an **acute angle**.

Acute angles are smaller than right angles. DÊF is an acute angle.

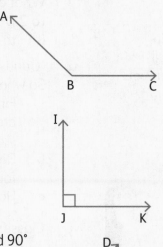

Maths ideas

In this unit you will:
* revisit points, lines and angles
* measure, classify and name angles
* identify parallel and perpendicular lines.

Key words

points	acute angle
lines	obtuse angle
angles	straight angle
line segment	protractor
vertex	scale
arms	perpendicular
right angle	parallel

Important symbols

AB̂C	angle ABC
└	right angle
⊥	perpendicular
//	parallel

An angle that measures between 90° and 180° is called an **obtuse angle**. Obtuse angles are bigger than right angles. NÔP is an obtuse angle.

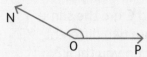

If you put two right angles together you get an angle that looks like a straight line. This is called a **straight angle** and it measures 180°. QR̂S is a straight angle.

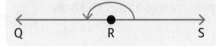

1   Three angles are marked on the diagram.
    a   Name each angle using three letters.
    b   Write the type of each angle next to its name.

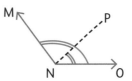

2   Without measuring, write down what type each of these angles is.

a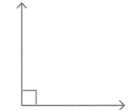

b

c

d

e

f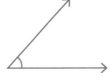

Explain

This is a **protractor**. It is used to measure the size of angles in degrees.

A protractor has two **scales**.

The inner scale goes from 0° to 180° from the right.
The outer scale goes from 0° to 180° from the left.

To measure the size of an angle:

Step 1: Put the centre point marked ✛ on the vertex of the angle.

Step 2: Line up any arm of the angle with the 0° line. You will use the scale that matches this zero.

Step 3: Find where the other arm of the angle lies on the correct scale.

This is the measurement of the angle in degrees.

The diagrams show how to do this.

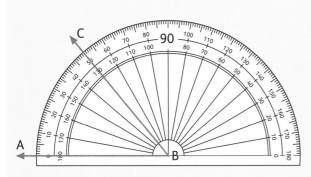

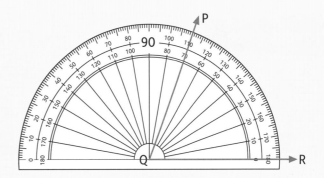

This angle measures 50°.
You read AB̂C on the outer scale, because the bottom arm is on zero on that scale.

PQ̂R measures 70°.
You read the size on the inner scale, because the bottom arm is on zero on that scale.

3   Measure each angle using a protractor. Write the name of each angle and its size.

a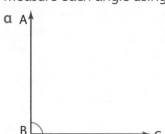

b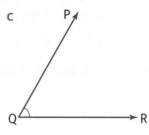

c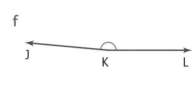

d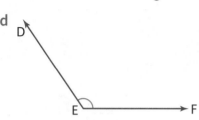

e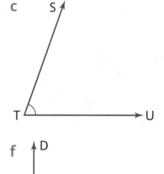

f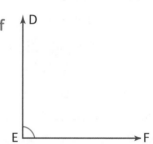

4   Write down an estimate of the size of each angle below.

a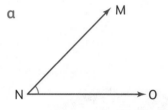

b

c

d

e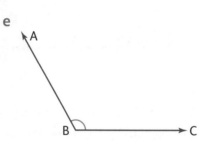

f

5   Now measure the angles in Question 2 using a protractor and write down their actual sizes.

6   Compare your answers with your partner's to see whose estimates were the closest. Talk about how you can estimate the size of angles more accurately.

Investigate

7   Where would you place your protractor to measure the size of each angle in this triangle?

Try out your ideas.
Write the size of each angle you measure.
Add the three sizes together.
If the angles don't add up to 180°, you have not measured accurately enough.

8   How can you use a protractor to draw an angle of 115°? Share your ideas and then draw the angle. Label it XŶZ.

Explain

Remember that lines that intersect or meet at right angles are **perpendicular**.

The right-angle symbol (⌐) shows that lines are perpendicular (⊥).

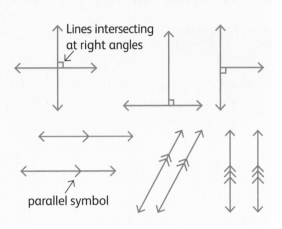

Lines intersecting at right angles

Parallel lines are the same distance apart all along their length. They run alongside each other and never meet or intersect.

The arrow symbols on the lines show they are parallel.

parallel symbol

9 This is a design for a phone cover. Work in pairs to complete the activities.

   a How many acute angles can you find in the design?

   b Estimate the size of the greatest and smallest acute angles. Use your protractor to measure the angles and see how well you estimated.

   c Find five obtuse angles in the design. Measure them.

   d How many pairs of parallel lines can you find? Compare your answers with another pair.

   e Find all the perpendicular lines in the design by identifying the right angles.

   f Design and draw your own geometric design for a phone cover. Include all the different types of angles as well as parallel and perpendicular lines. Describe it to your partner once you've finished.

What did you learn?

Draw and label the following:

1 acute angle $A\hat{B}C$

2 right angle $X\hat{Y}Z$

3 obtuse angle $P\hat{Q}R$

4 a line segment MN that is exactly 5 cm long

5 line segment AB perpendicular to line XY.

B Shapes and their properties

A **polygon** is a closed **plane shape** with at least 3 straight sides. Polygons are named according to the number of sides and angles they have.

Read the information in the table to make sure you remember these polygons.

Maths ideas

In this unit you will:
* revise the names and properties of 2-D shapes
* classify shapes using their properties
* identify and name different types of quadrilaterals.

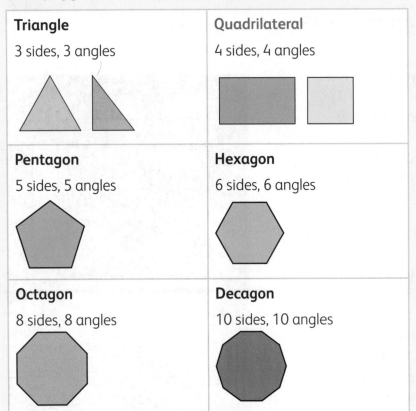

Triangle	Quadrilateral
3 sides, 3 angles	4 sides, 4 angles
Pentagon	**Hexagon**
5 sides, 5 angles	6 sides, 6 angles
Octagon	**Decagon**
8 sides, 8 angles	10 sides, 10 angles

Key words

polygon
plane shape
quadrilateral
circle
circumference
radius
diameter
square
rectangles
parallelograms
rhombus
trapezium
kite

A polygon is regular if all its sides are equal in length and all its angles are the same size.

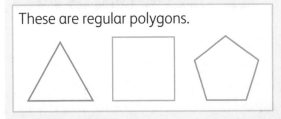

These are regular polygons.

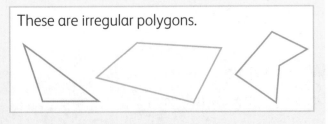

These are irregular polygons.

Circles are plane shapes, but they are not polygons as they do not have straight sides.

1 Write the name of each shape.

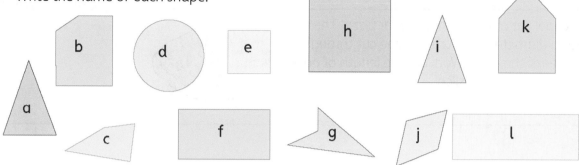

Explain

A circle is a closed plane shape with all points on its **circumference** the same distance from the centre.

The distance around the circle is called the circumference of the circle.

The **radius** of a circle is a line segment that runs from the centre of the circle to any point on the circumference of the circle. All the radii of a circle are the same length.

The **diameter** of a circle is the line segment that passes from a point on the circumference of the circle through the centre and to the opposite edge. The diameter divides a circle into two equal halves. The length of the diameter is twice the length of the radius. Wherever you draw the diameter of a circle, it will always be the same length.

2 Name the parts of the circle marked in red.

a
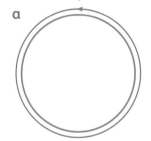
b
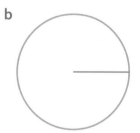
c
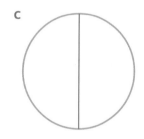
d
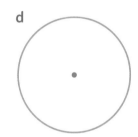

3 The radius of a circle is 2.5 cm. What is the diameter?

4 The centre point of a circle is 11 cm from its circumference. What is the radius of the circle?

5 Rodney runs a circular route three times a week. Today, he ran it five times and covered a distance of 2.5 km. What is the circumference of the circular route?

6 What type of angle is formed by the two radii of each circle?

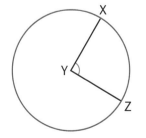
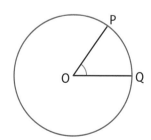
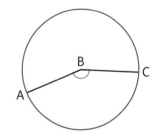
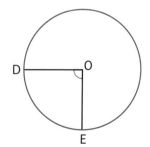

7   Rachael cut a slice from a circular pizza. The pizza has a diameter of 25 cm and she cut through the centre of the pizza. Work out the length of each straight side of the slice.

Explain

You already know that any polygon with four sides is a quadrilateral. We classify quadrilaterals into smaller groups using their properties.

Read the information and study the diagrams carefully.

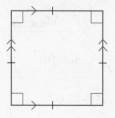

The only regular quadrilateral is a **square**.

Squares have four sides of equal length and four right angles.

The opposite sides are parallel to each other.

On the diagram, the small lines on each of the four sides show that they are equal. The small square in each angle indicates that it is a right angle. The parallel sides are marked with arrow pairs.

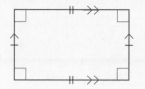

Rectangles have four right angles. The opposite sides of a rectangle are equal in length and parallel to each other.

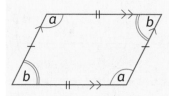

Parallelograms have opposite sides that are equal in length and parallel to each other. The opposite angles of a parallelogram are equal in size.

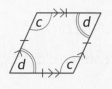

A **rhombus** has four equal sides with opposite sides parallel to each other and opposite angles equal.

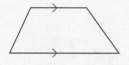

A **trapezium** has only one pair of parallel sides, which are not equal in length.

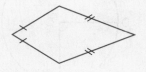

A **kite** has two pairs of equal sides that lie next to each other.

8   Use the properties of each quadrilateral to classify it.

a      b      c   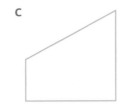   d      e

9   What properties do the quadrilaterals in each group have in common?

a

b

c

10 Which of these shapes are parallelograms?

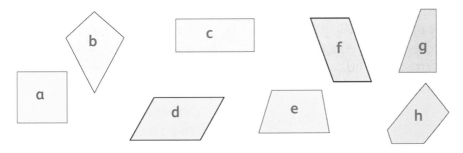

11 Say whether each statement is true or false. Give a reason for each answer.
    a   All squares are parallelograms.
    b   All parallelograms are squares.
    c   A rectangle is also a parallelogram.
    d   All parallelograms are rectangles.

What did you learn?

1   Draw a circle and label the diameter and circumference.

2   Write the names of quadrilaterals that have:
    a   two pairs of opposite sides parallel
    b   four sides of the same length
    c   opposite angles that are equal in size
    d   four right angles.

C Classifying triangles

Last year you classified triangles using the size of their angles.

An **acute-angled** triangle has three acute angles.

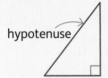

A **right-angled** triangle has one angle that measures 90°. The side opposite the right angle is always the longest side and is called the **hypotenuse**.

An **obtuse-angled** triangle has one obtuse angle.

Triangles can also be classified and named according to how many equal sides they have.

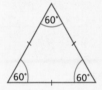

An **equilateral** triangle has three equal sides. All three angles are also equal and each measures 60°.

The equilateral triangle is the only regular three-sided polygon.

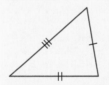

A **scalene** triangle has no equal sides and no equal angles.

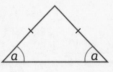

An **isosceles** triangle has two equal sides. The two angles at the base of the equal sides are also equal.

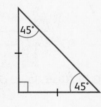

You can also name triangles using a combination of side and angle properties. This is a right-angled isosceles triangle. It has a right angle and two equal sides. Because the sides are equal, the two other angles are also equal.

Maths ideas

In this unit you will:
* revise what you know about triangle types
* learn to classify triangles using their side properties.

Key words

acute-angled

right-angled

hypotenuse

obtuse-angled

equilateral

isosceles

scalene

1   Draw any two triangles on a sheet of paper.
   a   Measure the sizes of the angles in each triangle.
   b   How does the length of the sides connect to the size of the angles?
   c   Add the three angle measurements. What do you notice?
   d   Compare your answers with others in your group.

2 Next draw a straight angle.
   a Cut out one of your triangles and cut or tear off the angles.
   b Place them next to each other with their vertices against the same point on the line.
   c What does this tell you about the sum of angles in the triangle?

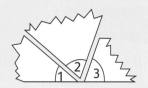

3 Name each type of triangle as accurately as you can.

   a 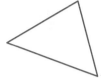 **b**   **c**   **d**   **e**

4 Use the properties of different types of triangles to answer these questions.

   a What is the length of the third side in this equilateral triangle?

5 cm   5 cm   ?

   b What is the size of AB̂C?

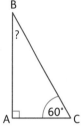

   c How can you tell which side of KL̂M is the longest and which side is the shortest without measuring?

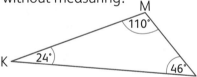

   d What is the length of QS?

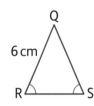

   e Write down any three possible measurements for the angles of an acute-angled scalene triangle.

Challenge

5 The sizes of two angles of each triangle are given. Work out the size of the third angle and classify each triangle as accurately as you can.
   a 10°   40°   **b** 40°   50°   **c** 120°   30°   **d** 43°   38°

What did you learn?

1 What type of triangle has three angles each measuring 60°?

2 What is the sum of the other two angles of a right-angled triangle?

3 Can an obtuse-angled triangle also be an isosceles triangle?

4 Is it possible to have more than one right angle in a triangle? Explain your answer.

D Shapes around us

If you build shapes with straws or strips and push them from one side, you will see that triangles are **stable**, but quadrilaterals are not.

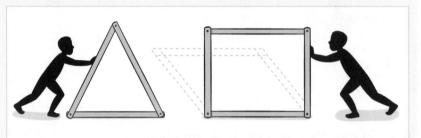

This is one of the main reasons why triangles are used in daily life to make structures strong and stable. For example, bridge supports are often triangular to make the bridge stronger and to stop it from bending or moving when a heavy load is placed on it.

In this unit you will:
* investigate how triangles are used in different structures.

stable

1 Find the triangles in each object. Talk about how they make each object stronger and more stable.

a

b

c

Work in pairs to find examples of triangles used in structures in your community. Take photographs if possible. You could also sketch the structure and highlight the triangles.

Make a poster showing what you find.

Choose one of the structures and write a few sentences to explain how the triangles make it safer, stronger and more stable.

Your teacher may choose a few students to present their posters and information to the class.

Topic 4 Review

Key ideas and concepts

Copy the mind map. Draw diagrams and write short notes to summarise the main points covered in each unit in this topic.

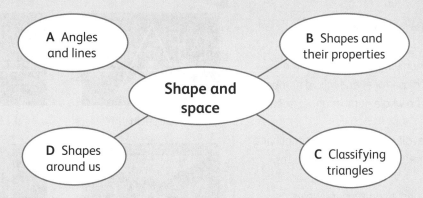

Think, talk, write …

1   Why do you think it is important to know the correct mathematical names for shapes and parts of shapes? Share your ideas with your group.

2   Find one shape pattern at home or in the community. Photograph it or draw it and paste it into your maths journal. Write a short description of the pattern using mathematical language.

Quick check

1   What do you call an angle that is greater than a right angle, but smaller than a straight angle?

2   Which quadrilaterals have four right angles?

3   Is a trapezium a special type of parallelogram? Explain why or why not.

4   Can an acute-angled triangle also be scalene, isosceles or equilateral?

5   How can you work out the length of the radius of a circle if you have the length of its diameter?

6   Identify as many different shapes as you can in the diagram. Use letters to name each shape you find and write the type of shape next to each one.

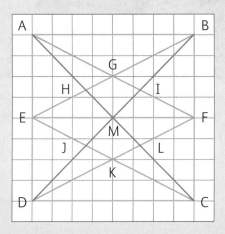

Number sense (2)

Teaching notes

Fractions

* Students should be familiar with common fractions (such as $\frac{2}{3}$, $\frac{7}{8}$ and $\frac{9}{10}$), improper fractions (with a numerator greater than the denominator, such as $\frac{11}{7}$ or $\frac{9}{4}$) and mixed numbers (a combination of a whole number and a fraction, such as $1\frac{1}{2}$ and $3\frac{3}{4}$).

* In this topic we revisit equivalence and simplifying fractions and convert between different forms of fractions.

* Remind students about common multiples, as they will use these to find the lowest common denominator (LCD) of a set of fractions so that they can write them with the same denominator to compare and order them.

Decimals

* Decimals (tenths and hundredths) can be indicated on the place value table by extending it to the right. Make sure students understand that everything to the right of the ones place is 'less than 1'.

* Revise the rules for rounding numbers with the class if necessary before they have to round off decimals. Make sure they realise that rounding to one decimal place means rounding to tenths (and vice versa).

Percentages

* A percentage is a fraction with a denominator of 100. It is written using a per cent symbol (%) rather than as a fraction, so $\frac{45}{100}$ would be written as 45%.

* Students need to be able to convert between percentages, decimals and fractions and they will focus on doing that in Unit C.

Ratio and proportion

* A ratio is a comparison of two or more quantities measured in the same units. Mixing quantities such as 1 cup water to 3 cups flour can be written as a ratio (1 : 3). We read this as 1 to 3. This is not the same as $\frac{1}{3}$ because 1 cup water plus 3 cups flour makes 4 cups, so the water is $\frac{1}{4}$ of the mixture!

* Ratios can be simplified by dividing all number parts of the ratio by the same number.

St Lucia   Barbados

Grenada   St Vincent and the Grenadines

Estimate what fraction of each flag is yellow. Tell your partner how you got your answers. Which estimate was easiest? Why? Which was the most difficult to estimate? Why?

Identify the decimals in the photograph. What does each one mean? Why do these decimals only go to hundredths? When might a price not have a decimal part?

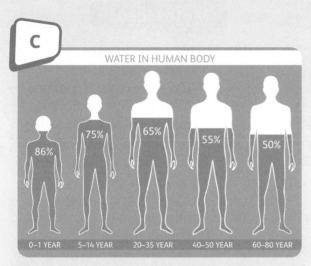

WATER IN HUMAN BODY

86% — 0–1 YEAR
75% — 5–14 YEAR
65% — 20–35 YEAR
55% — 40–50 YEAR
50% — 60–80 YEAR

What does this poster tell you? What fraction of your body is composed of water? What fraction of a 70-year-old person's body is water? How could you express the amount of water in a 1-year-old as a fraction and as a decimal?

The instructions for making concrete say that you need to mix 1 part cement, 3 parts sand and 3 parts stone. What does this mean? How much stone will you need if you have $\frac{1}{2}$ a bag of cement? What could happen if you don't mix the right amount of materials?

Think, talk and write

A Revisit fractions *(pages 46–49)*

1 A school flag is $\frac{1}{4}$ blue, $\frac{2}{8}$ green and $\frac{3}{12}$ red.
   The rest is white. What fraction is white?

2 a Which is greater, $\frac{2}{3}$ or $\frac{3}{10}$?

   b Which is smaller, $\frac{7}{20}$ or $\frac{2}{3}$?

3 Draw your own diagrams to show the fractions $\frac{7}{3}$ and $2\frac{1}{3}$.

B Revisit decimals *(pages 50–51)*

1 a What digit is in the hundredths place in the number 88.72?

   b What digit is in the tenths place in the number 73.01?

2 a Write the steps you would follow to round 14.67 to the nearest whole number.

   b Give two examples of when a rounded decimal value is good enough to use.

   c Give two examples of where you would need to use the exact amount and not a rounded value.

C Fractions, decimals and percentages *(pages 52–53)*

Write notes to explain how fractions, decimals and percentages are similar and how they differ.

D Ratio and proportion *(pages 54–56)*

1 a Explain what this instruction on a bottle of concentrated juice means: Dilute the concentrated juice as follows: 1 part of juice to 5 parts of water.

   b How much water should you add if you have:
      i   2 cups of concentrate?
      ii   30 ml of concentrate?

A Revisit fractions

A **fraction** is a number of **parts** of a **whole**.

The **denominator** tells you how many equal parts the whole is divided into.

The **numerator** tells you how many of the parts you are working with.

numerator
denominator $\dfrac{3}{4}$ 3 out of 4 equal parts

Mixed numbers have a whole number part and a fraction part.

They can be regrouped to have a numerator that is greater than a denominator to make them easier to work with.

mixed number

$3\dfrac{1}{7}$

whole number   fraction

1 whole $= \dfrac{7}{7}$

3 wholes $= \dfrac{7}{7} + \dfrac{7}{7} + \dfrac{7}{7} = \dfrac{21}{7}$

So, $3\dfrac{1}{7} = \dfrac{21}{7} + \dfrac{1}{7} = \dfrac{22}{7}$

Equivalent fractions have the same value.

Equivalent fractions can be found by multiplying or dividing the numerator and the denominator by the same number.

$\dfrac{1}{4} \times \dfrac{3}{3} = \dfrac{3}{12}$, so $\dfrac{1}{4} = \dfrac{3}{12}$
$\qquad$
$\dfrac{7}{9} \times \dfrac{4}{4} = \dfrac{28}{36}$, so $\dfrac{7}{9} = \dfrac{28}{36}$

$\dfrac{6}{18} \div \dfrac{6}{6} = \dfrac{1}{3}$, so $\dfrac{6}{18} = \dfrac{1}{3}$
$\qquad$
$\dfrac{30}{50} \div \dfrac{10}{10} = \dfrac{3}{5}$, so $\dfrac{30}{50} = \dfrac{3}{5}$

$\dfrac{1}{3}$ is the simplest form of $\dfrac{6}{18}$ and $\dfrac{3}{5}$ is the simplest form of $\dfrac{30}{50}$.

A fraction is in simplest form when no whole number except 1 can divide into both the numerator and denominator.

To **simplify** (find the simplest form of) any fraction you divide the numerator and denominator by the highest common factor of both numbers. There are different ways to do this.

In this unit you will:
* revise what you learnt in earlier levels about common fractions, improper fractions and mixed numbers
* convert fractions from one type to another
* generate equivalent fractions and simplify fractions
* find a common denominator and use it to arrange fractions by size.

fraction
parts
whole
denominator
numerator
mixed numbers
equivalent
simplify
improper fraction
compare
order
lowest common denominator

Example

Write $\dfrac{16}{40}$ in simplest form.

Method 1

$\dfrac{16}{40} \div \dfrac{8}{8} = \dfrac{2}{5}$

Divide by the HCF = 8.

Method 2

$\dfrac{16}{40} \div \dfrac{4}{4} = \dfrac{4}{10}$

$\dfrac{4}{10} \div \dfrac{2}{2} = \dfrac{2}{5}$

Work in steps.

Method 3

$\dfrac{16^2}{40_5} = \dfrac{2}{5}$

Use cancelling (a short way of writing the division).

1 Write each **improper fraction** as a mixed number.

a $\frac{11}{3}$ b $\frac{19}{3}$ c $\frac{145}{12}$ d $\frac{94}{7}$

2 Write each mixed number as an improper fraction.

a $3\frac{3}{4}$ b $5\frac{1}{2}$ c $8\frac{1}{4}$ d $3\frac{7}{9}$

3 Are these fractions equivalent? Write yes or no.

a $\frac{8}{12}$ and $\frac{2}{3}$ b $\frac{3}{10}$ and $\frac{3}{5}$ c $\frac{4}{12}$ and $\frac{1}{4}$

4 Divide the numerator and denominator of each fraction by the same number and write the fraction in simplest form.

a $\frac{56}{84}$ b $\frac{33}{99}$ c $\frac{42}{48}$ d $\frac{75}{100}$ e $\frac{84}{96}$

5 Write each fraction as an equivalent fraction with a denominator of 10.

a $\frac{1}{2}$ b $\frac{6}{20}$ c $\frac{12}{30}$ d $\frac{4}{40}$ e $\frac{1}{5}$

6 Write each fraction as an equivalent fraction with a denominator of 100.

a $\frac{1}{2}$ b $\frac{1}{4}$ c $\frac{3}{4}$ d $\frac{1}{5}$ e $\frac{1}{10}$

f $\frac{3}{10}$ g $\frac{1}{20}$ h $\frac{1}{25}$ i $\frac{12}{50}$ j $\frac{25}{50}$

7 Write each fraction in simplest form.

a $\frac{12}{32}$ b $\frac{15}{45}$ c $\frac{42}{49}$ d $\frac{36}{42}$ e $\frac{80}{100}$

f $\frac{6}{10}$ g $\frac{27}{72}$ h $\frac{11}{33}$ i $\frac{18}{60}$ j $\frac{35}{45}$

8 Find and write down all the sets of equivalent fractions in the box. Circle the fraction in each set that is in simplest form.

| $\frac{1}{5}$ | $\frac{10}{16}$ | $\frac{75}{100}$ | $\frac{2}{7}$ | $\frac{10}{50}$ | $\frac{15}{24}$ | $\frac{3}{4}$ |
| $\frac{10}{14}$ | $\frac{6}{30}$ | $\frac{5}{8}$ | $\frac{6}{8}$ | $\frac{12}{16}$ | $\frac{3}{15}$ | $\frac{25}{400}$ |

9 How many equivalent fractions with a smaller denominator can you make for each fraction?

a $\frac{160}{200}$ b $\frac{42}{63}$ c $\frac{42}{56}$

Problem-solving

10 A fraction is equivalent to $\frac{2}{3}$ and the sum of its numerator and denominator is 15. What is the fraction?

Ordering and comparing fractions

Explain

You can use the signs $<$, $=$ or $>$ to **compare** two fractions.

To compare fractions with the same denominators, look at the numerators.

$$\frac{1}{8} < \frac{5}{8} \qquad \frac{7}{9} > \frac{5}{9} \qquad \frac{17}{40} < \frac{30}{40}$$

To compare fractions with different denominators, use equivalent fractions.

You can write sets of fractions in ascending or descending **order**.

To order sets of fractions, write equivalent fractions with the same denominator.

It is more efficient to work with fractions if you use the **lowest common denominator** (LCD) of the fractions you are working with. The LCD is the lowest common multiple (LCM) of the denominators.

Example

Compare $\frac{2}{3}$ and $\frac{3}{5}$.

$\frac{2}{3} \times \frac{5}{5} = \frac{10}{15}$ Change both fractions to get fifteenths

$\frac{3}{5} \times \frac{3}{3} = \frac{9}{15}$

$\frac{10}{15} > \frac{9}{15}$, so $\frac{2}{3} > \frac{3}{5}$

Example

Arrange these fractions in ascending order.

$$\frac{3}{5} \qquad \frac{3}{4} \qquad \frac{7}{10} \qquad \frac{1}{2}$$

Look at the denominators: 5, 4 10 and 2.

All the denominators are factors of 20.

Change the fractions to make equivalent twentieths.

$$\frac{3}{5} \times \frac{4}{4} = \frac{12}{20} \qquad\qquad \frac{3}{4} \times \frac{5}{5} = \frac{15}{20}$$

$$\frac{7}{10} \times \frac{2}{2} = \frac{14}{20} \qquad\qquad \frac{1}{2} \times \frac{10}{10} = \frac{10}{20}$$

The order is $\frac{10}{20}, \frac{12}{20}, \frac{14}{20}, \frac{15}{20}$

Now you can write the original fractions in ascending order.

$$\frac{1}{2} \qquad \frac{3}{5} \qquad \frac{7}{10} \qquad \frac{3}{4}$$

These methods work for mixed numbers too. But remember to compare the whole numbers first – it does not matter what the fraction part is if the whole numbers are different.

Example

Compare $2\frac{1}{2}$ and $3\frac{3}{4}$.

Look at the whole numbers:

$3 > 2$, so $3\frac{3}{4} > 2\frac{1}{2}$

1   Which fraction is greater in each pair?

   a $\frac{2}{7}$ or $\frac{3}{8}$ 　　　　　 b $\frac{9}{10}$ or $\frac{6}{10}$ 　　　　　 c $\frac{11}{100}$ or $\frac{17}{100}$

   d $\frac{4}{5}$ or $\frac{9}{10}$ 　　　　　 e $\frac{5}{6}$ or $\frac{11}{12}$ 　　　　　 f $\frac{6}{11}$ or $\frac{22}{33}$

2   Rewrite each fraction with a denominator of 48 and then arrange the original fractions in ascending order.

$$\frac{5}{6} \qquad \frac{1}{3} \qquad \frac{11}{12} \qquad \frac{19}{24} \qquad \frac{3}{4} \qquad \frac{5}{8} \qquad \frac{9}{24} \qquad \frac{21}{24}$$

3   Find the LCD of each set of fractions and then arrange the fractions in descending order.

a $\frac{2}{15}$ $\frac{3}{5}$ $\frac{1}{10}$

b $\frac{2}{3}$ $\frac{4}{9}$ $\frac{11}{27}$

c $\frac{1}{3}$ $\frac{3}{4}$ $\frac{3}{8}$ $\frac{5}{12}$

d $\frac{5}{12}$ $\frac{3}{4}$ $\frac{13}{56}$

e $\frac{3}{10}$ $\frac{1}{2}$ $\frac{3}{4}$ $\frac{2}{5}$ $\frac{1}{4}$

f $\frac{1}{2}$ $\frac{1}{4}$ $\frac{3}{10}$ $\frac{31}{100}$ $\frac{2}{5}$

Problem-solving

4   The number line shows the position of some fractions.

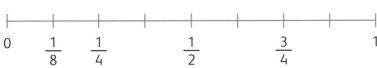

0     $\frac{1}{8}$ $\frac{1}{4}$ $\frac{1}{2}$ $\frac{3}{4}$ 1

Between which two values on the number line would you place these fractions?

a $\frac{3}{16}$ b $\frac{1}{10}$ c $\frac{7}{16}$ d $\frac{5}{8}$ e $\frac{7}{10}$ f $\frac{28}{32}$

5   Each sack had a mass of 1 kilogram when it was full. The mass of each one now is written on it as a fraction of a kilogram.

a   Write the masses in order from lightest to heaviest.   b   Which sacks are more than half full?

 $\frac{3}{8}$ kg $\frac{4}{15}$ kg $\frac{3}{5}$ kg $\frac{1}{3}$ kg $\frac{7}{12}$ kg $\frac{2}{3}$ kg

6   The amount of homework that each student has completed is given below.

Jayden $\frac{5}{6}$	Micah $\frac{7}{8}$	Sharyn $\frac{6}{7}$	Shayna $\frac{4}{5}$	Dezi $\frac{9}{10}$

a   Write their names in order from most homework completed to least homework completed.

b   Which students have completed more than half their homework?

What did you learn?

1   Four fractions are equivalent in each set. Find the one that is not equivalent and write it in simplest form.

Set A: $\frac{9}{21}$ $\frac{18}{42}$ $\frac{15}{27}$ $\frac{3}{7}$ $\frac{6}{14}$ Set B: $\frac{42}{60}$ $\frac{56}{70}$ $\frac{21}{30}$ $\frac{14}{20}$ $\frac{7}{10}$

2   Jayden and Alex have the same amount of homework. Jayden has completed $\frac{11}{15}$ of his homework and Alex has completed $\frac{4}{5}$ of his. Who has done the most homework?

3   Which fraction is smaller in each pair?

a $\frac{3}{4}$ or $\frac{2}{3}$ b $\frac{1}{4}$ or $\frac{5}{12}$ c $\frac{2}{3}$ or $\frac{11}{12}$

4   Find the LCD and write these fractions in ascending order.

$\frac{1}{4}$ $\frac{6}{10}$ $\frac{2}{3}$ $\frac{1}{5}$ $\frac{5}{6}$ $\frac{7}{15}$

B Revisit decimals

Decimals are numbers that use a **decimal point** and **place value** to show fractions. The number to the left of the decimal point is a whole number, while the numbers to the right show a number that is smaller than 1 (a fraction).

The number 43.72 consists of the following:
* 43 (a whole number)
* 7 **tenths** ($\frac{7}{10}$)
* 2 **hundredths** ($\frac{2}{100}$)

We can show the decimal number on the place value table like this:

Thousands	Hundreds	Tens	Units	.	tenths	hundredths
		4	3	.	7	2

When you read this number, you say forty-three point seventy-two.

As a mixed number, it would look like this: $43\frac{72}{100}$, which is $43\frac{18}{25}$ in simplest form.

You can also write the number in expanded form, like this:

$43.72 = 40 + 3 + \frac{7}{10} + \frac{2}{100}$

To compare decimals, you line up the decimal points and then compare the **digits** in each place, starting from the left.

Maths ideas

In this unit you will:
* revise what you learnt about decimals and decimal place value in earlier levels
* compare and order decimals
* round decimals to a given place.

Key words

decimals
decimal point
place value
tenths
hundredths
digits

Example 1

Which is greater: 1.87 or 1.9?
  8 < 9, so 1.87 < 1.9

1.87 — You can fill in 0 to show
1.90   there are no hundredths

Line up the decimal points

Think: 1 and 87 hundredths is smaller than 1 and 90 hundredths.

You can round decimals to the nearest tenth or whole number using the same rules you use for whole numbers.

Rounding to the nearest tenth is also called rounding to one decimal place.

Example 2

Round 4.53 to the nearest whole number and to the nearest tenth.

Look at the digit to the right of the rounding place.

4.53 ↓

5 or more, so

4.53 is 5 to the nearest whole number.

4.53 ↓

3 is less than 5, so

4.53 is 4.5 to the nearest tenth.

1   These numbers all need a decimal point. Decide where the decimal point should go to make each value realistic.

   a   The top speed of a cheetah   1 158 km/h

   b   The price of a pair of running shoes $4 595

   c   The average mass of a 12-year-old boy   399 kg

   d   The test score of the top mathematics student   972%

   e   The average height of a newborn baby   4 955 cm

2   Use the symbols < and > to compare these decimals.

   a   23.07 ☐ 23.4   b   67.91 ☐ 67.92   c   7.74 ☐ 7.47

   d   1.99 ☐ 1.9   e   300.99 ☐ 300.09   f   3.6 ☐ 3.65

3   Round each decimal to the nearest whole number.

   a   805.6   b   6 095.1   c   3.8   d   20.09   e   674.3   f   19.9

4   Round each decimal to the nearest tenth.

   a   42.82   b   69.17   c   65.19   d   81.89   e   9.99   f   12.05

5   Four babies are born on the same day. Baby André has a mass of 3.37 kg, baby Benni has a mass of 3.38 kg, baby Colin has a mass of 3.73 kg and baby Denae has a mass of 3.52 kg. Write the babies' names in order from the heaviest baby to the lightest baby.

6   This drawing shows the heights of some trees at a nature reserve in Bridgetown. List the heights in order from tallest to shortest.

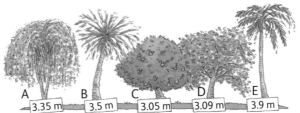

A 3.35 m   B 3.5 m   C 3.05 m   D 3.09 m   E 3.9 m

Problem-solving

7   In a javelin competition, each person gets six throws, but only the best three results count. These are Albert's results for six throws.

   84.58 m   84.39 m   85.28 m   82.67 m   85.38 m   84.38 m

   a   Arrange the results in order from least to greatest distance.

   b   Which three results would be counted in this competition?

   c   The Olympic record for men's javelin is 90.57 m (set in 2008). Look at Albert's best throw in this competition. How much further would he need to throw the javelin to beat the Olympic record?

   d   In the next competition, Albert throws $\frac{3}{10}$ of a metre further than his best in this competition. What is that distance?

What did you learn?

1   Write each set of decimals in ascending order.

   a   12.92, 13.58, 13.29, 13.51, 12.78

   b   0.23, 0.03, 0.20, 0.2, 0.3

2   Round each decimal in set a above to the nearest whole number and to the nearest tenth.

C Fractions, decimals and percentages

Explain

Per cent means 'for each hundred'. The symbol % is read as per cent and it shows that you are dealing with a **percentage**.

25% is read as twenty-five per cent.

One hundred per cent, or 100%, is the whole.

(Remember $\frac{100}{100}$ = 1.)

If you have 70%, it means you have $\frac{70}{100}$ or 0.7 of the whole.

It is possible to have more than 100%. If you have 150%, it means you have the whole plus an additional half, similar to the mixed number $1\frac{1}{2}$.

Look at these examples to see where you might find percentages greater than 100% in daily life.

Tourist numbers 112% of last year	Rainfall during hurricane season was 120% of average.	Each tablet provides 115% of your daily vitamin C allowance.
Charity raises 166% of their funding goal.	*My new salary is 150% of my old one.*	Damian sold 114% of his quota.

Maths ideas

In this unit you will:
* revise the concept of per cent
* explain how percentages are used in daily life
* convert between equivalent fractions, decimals and percentages.

Key words

per cent

percentage

convert

Fractions, decimals and percentages are really just different ways of describing the same thing and you can **convert** between them.

Look at this number line:

0% 10%	25% 35%	50%	75%	90% 100%
0 $\frac{1}{10}$	$\frac{1}{4}$ $\frac{7}{20}$	$\frac{1}{2}$	$\frac{3}{4}$	$\frac{9}{10}$ 1
0 0.1	0.25 0.35	0.5	0.75	0.9 1

Can you see how to convert between percentages and decimals?

$12\% = \frac{12}{100} = 0.12$ $0.76 = \frac{76}{100} = 76\%$

To convert percentages to fractions, write the percentage as a fraction with a denominator of 100 and simplify if you can.

$25\% = \frac{25}{100} = \frac{1}{4}$ $35\% = \frac{35}{100} = \frac{7}{20}$

To write fractions as percentages, convert the fraction to an equivalent fraction with a denominator of 100 and write that as a percentage.

$\frac{3}{4} = \frac{3}{4} \times \frac{25}{25} = \frac{75}{100} = 75\%$ $\frac{18}{20} = \frac{18}{20} \times \frac{5}{5} = \frac{90}{100} = 90\%$

1   Write each percentage as an equivalent fraction in simplest form.
   a 20%   b 50%   c 80%   d 92%   e 12%

2   Write each fraction as a percentage.
   a $\frac{20}{100}$ b $\frac{9}{10}$ c $\frac{6}{25}$ d $\frac{7}{50}$

   e $\frac{3}{5}$ f $\frac{9}{20}$ g $\frac{5}{8}$ h $\frac{19}{50}$

3   Write each percentage as a decimal.
   a 12%   b 98%   c 2%
   d 80%   e 7%   f 180%

4   Write each decimal as percentage.
   a 0.13   b 0.09   c 0.65   d 0.99   e 0.6
   f 1.5   g 2.5   h 0.42   i 0.01   j 1.0

5   Can you find your way through the maze? Check each number sentence to see if it is true (T) or
     false (F) and follow the correct arrow.

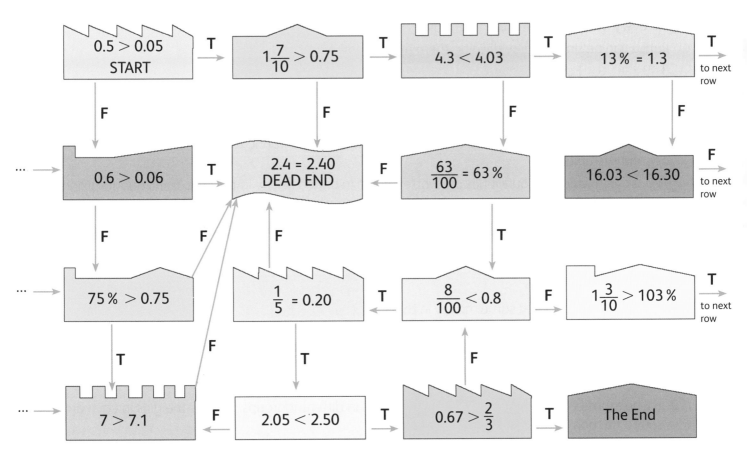

What did you learn?

Write each amount as a fraction in simplest form.

1   10%   2   0.85   3   1.25   4   150%   5   75%

D Ratio and proportion

Explain

Ratios can be used to compare two sets of data or numbers. Ratios, like fractions, show relationships between numbers, sets or parts of sets. You can write ratios and fractions in different ways.

There are 3 boys in a room. There are 9 girls in a room. The total number of people in the room is 12.

We can express these amounts as fractions, decimals or percentages. $\frac{1}{4}$, 0.25 or 25 % of the people in the room are boys and $\frac{3}{4}$, 0.75 or 75 % of the people in the room are girls.

The ratio of girls to boys in this room is nine to three (9 : 3).

For every nine girls there are three boys.

The ratio of boys to girls in this room is three to nine (3 : 9).

For every three boys there are nine girls.

Ratios can be simplified by dividing them by the same number.

$3 : 9 = \frac{3}{3} : \frac{9}{3} = 1 : 3$

3 : 9 and 1 : 3 are equivalent ratios.

Ratios do not have units. This is because the **proportions** are fixed, so the ratio works for all units.

A ratio of 1 : 2 could mean 1 cup to 2 cups, 1 bag to 2 bags, 1 tonne to 2 tonnes, or any other appropriate units.

The order in which the quantities are written down in a ratio is very important. Think about mixing orange juice from a concentrate. If the instructions say 1 part concentrate to 3 parts water, this is a ratio of 1 : 3.

This means that if you pour 1 cup of orange concentrate you have to add 3 cups of water. This will give you the correct mix and a pleasant drink.

Maths ideas

In this unit you will:
* revise how to use ratios to compare numbers and amounts
* express ratios in different ways.

Key words

ratios

proportions

1 Jay and Dean shared some money in the ratio 2 : 1.
   a If Jay got $40, how much did Dean get?
   b If Dean got $16, how much did Jay get?
   c If the total amount to be shared was $45, how much would they each get?

2 The ratios below show the ratio of girls to boys in different groups. Express the girls in each group as a fraction of the group in simplest form.
   a 6 : 18   b 10 : 25   c 20 : 35   d 24 : 36

3 Write the following ratios in their simplest form (as equivalent ratios).
   a 4 : 6   b 75 : 100   c 20 cm : 80 cm   d 4 : 40

4 Write five equivalent ratios for each of these ratios.
   a 2 : 5   b 1 : 2   c 4 : 7   d 7 : 8   e 4 : 5

5 Write each of these ratios as an equivalent ratio in simplest form.

    a 5 : 15   **b** 8 : 12   **c** 4 : 16   **d** 18 : 24   **e** 24 : 36

6 Write a ratio to represent each situation.

    a To make jam, add six kilograms of sugar for every three kilograms of fruit.

    b To make brass you need to mix copper and zinc in the following ratio: 65 : 35.

    c In a spoon made of pewter, 20 % of the pewter is silver and 80 % is tin.

    d For every 15 minutes that I sit down, I spend 1 hour walking around.

7 There are 3 girls for every 2 boys at a party.

    a Draw a diagram to show the ratio of girls to boys.

    b If there are 20 boys at the party, how many girls are there?

    c If there are 40 children at the party, how many are boys?

8 A recipe for 20 muffins uses 2 eggs. How many eggs would you need for:

    a 40 muffins?   **b** 10 muffins?   **c** 50 muffins?

Problem-solving

9 You can increase or decrease the size of a picture using a photocopier. Most copiers enlarge or reduce by a percentage.

Marita has copied a page at 75 %. This means her copy is 75 % or $\frac{3}{4}$ of the size of the original. The size of the copy compared to the original can be expressed as the ratio 75 : 100 or 3 : 4.

Write the ratio of the copy to the original if she reduces it by:

    a 50 %   **b** 20 %   **c** 80 %.

10 A class collects magazines to take to a nearby hospital. For every two magazines that go into the small collection box, four go into the big collection box.

    a Draw a diagram to show how the magazines are shared between the two collection boxes.

    b If 120 magazines were collected in all, how many went into each box?

11 Janine and Shay went mango picking. Janine picked 11 mangoes each time Shay picked 7 mangoes. Altogether they picked 54 mangoes. How many did each person pick?

12 Kershawn and Jay shared 51 mangoes. Kershawn got twice as many as Jay.
   a Draw boxes to show each person's share.
   b What was Kershawn's share?
   c How many mangoes did Jay get?

Challenge

13 Design a new rectangular flag for your school's eco-club.

   The sides of the rectangle are in the ratio 2 : 3.

   The flag should have three colours. The club wants to promote care of the environment, which includes the land and the ocean. Choose three appropriate colours and design the flag.

   Write down a suitable size for the flag and the ratios of the colours in the flag.

   Then use the ratios to calculate how much fabric of each colour you will need to make the flag.

What did you learn?

1 The table gives information about the number of boys and girls in Levels 5 and 6 at a school.

	Girls	Boys
Level 5	36	44
Level 6	40	32

Write these ratios in simplest form.
   a Level 5 girls to Level 6 girls
   b Level 5 boys to Level 6 boys
   c Level 5 students to Level 6 students
   d Boys to girls in both levels

2 Red and blue paint is mixed in the ratio 3 : 5 to produce purple paint. How much paint of each colour would you need to make 400 millilitres of purple paint?

3 A student collects $80 for charity. He shares the money between two charities in the ratio 2 : 3. How much money does he give to each charity?

Topic 5 Review

Key ideas and concepts

Copy this table and complete it to summarise what you learnt in this topic.

	Fractions	Decimals	Percentages	Ratios
Explain the concept and give examples.				
How do you compare or order them?				
How do you make equivalent forms?				
Can you simplify them? If so, how?				

Think, talk, write …

1 A friend asks you how to order a set of numbers that contains fractions, decimals and percentages. What would you tell them?

2 Where do you see fractions, decimals, percentages and ratios in everyday life? Make a list.

Quick check

1 One fraction in each set is not equivalent to the others. Which one is it?

a $\frac{4}{18}$ $\frac{8}{36}$ $\frac{14}{63}$ $\frac{2}{9}$ $\frac{32}{108}$ b $\frac{3}{4}$ $\frac{10}{12}$ $\frac{5}{6}$ $\frac{15}{18}$ $\frac{45}{54}$

2 Write each fraction as a decimal and an equivalent percentage.

a $\frac{7}{10}$ b $1\frac{3}{10}$ c $\frac{1}{2}$ d $\frac{4}{5}$

3 Copy the numbers. Insert a decimal point so that the 3 in each number has a value of three hundredths. You may need to add zeros as placeholders.

a 143   b 83   c 4593   d 13   e 403   f 3

4 Which is further, $\frac{65}{75}$ kilometres or 0.9 kilometres?

5 I have a fraction card. The fraction on the card is equivalent to $\frac{1}{3}$. If I subtract 3 from the numerator and 6 from the denominator, the new fraction is half the value of the one I started with. What is the fraction on the card?

6 Convert these marks to percentages and write them in descending order.

40 out of 80   90 out of 150   24 out of 30   70 out of 100

7 Dieticians recommend that a healthy diet should consist of half vegetables and fruit, a quarter proteins and a quarter carbohydrates. Which of these statements show the amount of vegetables we should eat in relation to the rest of the food on a plate.

$\frac{1}{2}$ 1 : 1   1 : 2   0.25   50%

Teaching notes

Adding and subtracting fractions

* Students already know how to add and subtract fractions with the same denominators and they have learnt to make equivalent fractions.

* When you work with fractions with different denominators, you use equivalence to find the lowest common denominator. This is simply the lowest common multiple of the denominators. If you don't use the lowest common denominator, you have to simplify after calculating, but you can still add and subtract the fractions.

* Mixed numbers are easier to work with if you convert them to improper fractions first. Then they are treated like any other fraction when you add or subtract.

Multiplying and dividing fractions

* To multiply a whole number by a fraction, you write it as a fraction with a denominator of 1 ($5 = \frac{5}{1}$, for example). Then you can multiply numerators by numerators and denominators by denominators. This is the same principle as when you multiply a fraction by any other fraction.

* The reciprocal of a number is the result you get when you divide 1 by that number, so any number multiplied by its reciprocal has a product of 1. The reciprocal of a fraction $\frac{a}{b}$ is $\frac{b}{a}$. For example, the reciprocal of $\frac{2}{5}$ is $\frac{5}{2}$ and $\frac{2}{5} \times \frac{5}{2} = \frac{10}{10} = 1$.

* When you divide a fraction by a whole number, such as $\frac{1}{2} \div 6$, you are really finding $\frac{1}{6}$ of the half. $\frac{1}{6}$ of a half means $\frac{1}{6} \times \frac{1}{2}$. This is why you can 'invert' the fraction you are dividing by ($\frac{6}{1}$ becomes $\frac{1}{6}$) and then multiply.

Operations with decimals

* Place value is important when working with decimals. Students need to line up decimal places and insert decimal points in products and in the quotient when they divide. Decimal calculations are otherwise done in the same way as whole number calculations.

A

Look at the photo. How can you describe the cut piece of apple mathematically? What is the mixed number that describes the whole set of apples? How many quarters are there in this amount of apples? How many apples would you have if you added another $1\frac{1}{4}$ apples to this amount?

B

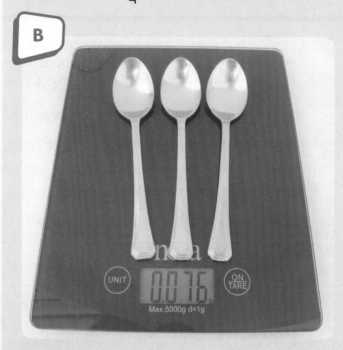

Read the mass of these three teaspoons on the scale display. What operation would you need to do to find the mass of one teaspoon? What would the scale display if another three teaspoons were added to it?

Think, talk and write

A Calculating with fractions *(pages 60–65)*

1. a How many metres of ribbon are there in all?

    $\frac{7}{12}$ m $\frac{1}{3}$ m $\frac{7}{9}$ m

    b How much do the two bags weigh in total?

    $2\frac{3}{5}$ kg $3\frac{9}{10}$ kg

    c Sandra drank $\frac{3}{5}$ of a litre of juice every day for four days. How much juice is this altogether?

    d Sherrie has three bars of chocolate. She gives $\frac{2}{5}$ of the chocolate to her sister. How much is $\frac{2}{5}$ of three bars?

Chocobar

Chocobar

Chocobar

B Calculating with decimals *(pages 66–69)*

1. The odometer in Ms Giffard's car showed 321.9 at the start of a journey.
    a What does that mean?
    b At the end of her journey, the odometer showed 537.6. How far did she travel?

2. Nadia wants to know the total length of these four pieces of ribbon.
    a Write the problem as a multiplication number sentence.
    b How many decimal places are in the number sentence?
    c How many decimal places are in the product?

    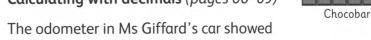

    0.12 m   0.12 m

    0.12 m   0.12 m

C Mixed problems *(page 70)*

1. Emily gave her grandmother $\frac{2}{5}$ of a loaf of bread.
    a What fraction of the bread was left?
    b If Emily's loaf had 20 slices, how many slices did Emily give her grandmother?

2. John gave 0.25 of a box of plums to his mother, and 0.2 to his friend. 0.3 of the plums were bad, so he threw them away and shared the rest with his sister.
    a What fraction of the plums did John give away?
    b What fraction of the plums did John end up with?

How do you think people who sell fabrics calculate with fractions and decimals? Make up two problems using lengths of fabrics and prices. Swap your problems with a partner. Solve the problems your partner set and then check each other's work and discuss the answers.

A Calculating with fractions

Explain

Do you remember how to add and subtract fractions and mixed numbers?

To add or subtract fractions, the denominators need to be the same.

When the **denominators** are different, you can convert them to **equivalent** fractions with the same denominator. You may need to regroup the answer if the numerator is greater than the denominator.

Maths ideas

In this unit you will:
* revise how to add, subtract and multiply fractions and mixed numbers
* learn how to divide fractions and mixed numbers by whole numbers
* solve problems involving calculations with fractions.

Key words

denominators   simplify
equivalent   reciprocal

Example 1

$\frac{3}{5} + \frac{9}{10} = \frac{6}{10} + \frac{9}{10} = \frac{15}{10}$

$\frac{15}{10}$ is fifteen tenths. This is equivalent to 1 whole ($\frac{10}{10}$) and $\frac{5}{10}$. We write this as the mixed number $1\frac{5}{10}$.
The $\frac{5}{10}$ part of the mixed number can be reduced to simplest form to get $\frac{1}{2}$.

Example 2

$1\frac{11}{12} + 1\frac{7}{12}$

$1 + 1 = 2$ Add the whole numbers

$\frac{11}{12} + \frac{7}{12} = \frac{18}{12}$ Add the fractions

$\frac{18}{12} = \frac{12}{12} + \frac{6}{12} = 1\frac{1}{2}$

$2 + 1\frac{1}{2} = 3\frac{1}{2}$

Example 3

$1\frac{1}{3} + 1\frac{5}{6} - 2\frac{1}{4}$

$= \frac{4}{3} + \frac{11}{6} - \frac{9}{4}$ Regroup the mixed numbers

$= \frac{16}{12} + \frac{22}{12} - \frac{27}{12}$ Convert to equivalent fractions with the same denominator

$= \frac{38}{12} - \frac{27}{12}$ Work from left to right
Add: $\frac{16}{12} + \frac{22}{12} = \frac{38}{12}$

$= \frac{11}{12}$ Subtract next: $\frac{38}{12} - \frac{27}{12} = \frac{11}{12}$

1 Calculate. Give answers in simplest form.

a $\frac{3}{5} + \frac{1}{5}$
b $\frac{7}{5} + \frac{3}{5}$
c $\frac{7}{8} - \frac{5}{8}$
d $\frac{11}{5} - \frac{4}{5}$

e $3\frac{14}{16} - 2\frac{11}{16}$
f $5\frac{9}{10} - 5\frac{7}{10}$
g $3\frac{7}{8} - 1\frac{7}{8}$
h $8\frac{3}{6} - 3\frac{5}{6}$

i $1 - \frac{3}{8}$
j $2 - 1\frac{1}{15}$
k $4 - \frac{5}{3}$
l $3 - \frac{3}{4}$

m $2\frac{3}{8} - \frac{5}{8}$
n $3\frac{2}{7} - 1\frac{5}{7}$
o $4\frac{2}{5} - 2\frac{4}{5}$
p $3\frac{1}{15} - 1\frac{9}{15}$

q $1\frac{4}{7} + \frac{5}{7}$
r $1\frac{9}{11} + \frac{9}{11}$
s $3\frac{7}{10} + 2\frac{9}{10}$
t $5\frac{2}{3} + 4\frac{2}{3}$

2   Calculate. Reduce your answers to simplest form if necessary.

a $\frac{3}{8} + \frac{3}{4}$ b $\frac{2}{3} + \frac{5}{6}$ c $\frac{3}{4} + \frac{9}{16}$ d $\frac{5}{8} + \frac{13}{16}$

e $\frac{9}{16} + \frac{1}{2}$ f $\frac{7}{10} + \frac{3}{5}$ g $\frac{7}{8} - \frac{3}{4}$ h $\frac{13}{16} - \frac{3}{4}$

i $\frac{3}{8} - \frac{1}{16}$ j $\frac{5}{7} - \frac{5}{14}$ k $\frac{17}{20} - \frac{3}{5}$ l $\frac{21}{25} - \frac{3}{5}$

3   Add. Give your answers in simplest form.

a $1\frac{3}{4} + 2\frac{1}{4}$ b $1\frac{1}{5} + 1\frac{2}{3}$ c $2\frac{1}{2} + 3\frac{1}{4}$ d $1\frac{1}{8} + 4\frac{1}{2}$

e $3\frac{7}{8} + \frac{5}{8}$ f $4\frac{3}{4} + 2\frac{1}{2}$ g $2\frac{5}{8} + 3\frac{1}{5}$ h $3\frac{7}{10} + 1\frac{2}{3}$

i $2\frac{7}{50} + 1\frac{19}{100}$ j $3\frac{3}{20} + 2\frac{3}{4}$ k $3\frac{2}{3} + 2\frac{4}{5}$ l $1\frac{2}{3} + 3\frac{4}{5}$

4   Subtract. Give your answers in simplest form.

a $4\frac{8}{10} - 1\frac{1}{2}$ b $3\frac{7}{8} - 1\frac{1}{4}$ c $4\frac{9}{10} - 2\frac{4}{5}$ d $3\frac{2}{3} - 1\frac{1}{4}$

e $5\frac{1}{2} - 3\frac{7}{8}$ f $4\frac{3}{5} - 1\frac{9}{10}$ g $5\frac{1}{4} - 2\frac{5}{12}$ h $4\frac{1}{2} - 2\frac{2}{3}$

i $4\frac{3}{10} - 2\frac{2}{3}$ j $4\frac{3}{8} - 1\frac{2}{5}$ k $1\frac{3}{8} - \frac{11}{16}$ l $7\frac{7}{15} - 4\frac{11}{12}$

5   Calculate.

a $\frac{3}{10} + \frac{2}{5} + \frac{4}{15}$ b $\frac{3}{4} + \frac{5}{6} + \frac{1}{2}$ c $\frac{3}{4} - \frac{1}{3} - \frac{1}{6}$ d $\frac{3}{4} + \frac{2}{5} - \frac{7}{10}$

e $\frac{2}{3} + \frac{3}{4} - \frac{1}{5}$ f $\frac{3}{4} - \frac{1}{3} + \frac{1}{2}$ g $2\frac{5}{9} + 2\frac{5}{6} - 2\frac{5}{18}$ h $4\frac{3}{4} - 1\frac{2}{3} + 3\frac{1}{2}$

i $2\frac{1}{4} + 3\frac{1}{6} - 3\frac{1}{8}$ j $3\frac{1}{5} + 2\frac{1}{15} - 2\frac{3}{4}$ k $2\frac{3}{8} - 1\frac{3}{16} + 3\frac{7}{8}$ l $2\frac{1}{4} - 1\frac{3}{5} + 3\frac{1}{2}$

Problem-solving

6   Try to solve these equations. Show your working.

a $x - 4\frac{1}{2} = 1\frac{1}{6}$ b $3\frac{1}{4} + x = 5\frac{19}{20}$ c $x - 2\frac{1}{4} = 3\frac{3}{5}$

7   Mrs Newton was baking. She mixed $3\frac{1}{2}$ cups of brown flour, $1\frac{2}{3}$ cups white flour, $\frac{2}{3}$ of a cup of bran and $\frac{1}{8}$ of a cup of sugar. How many cups is this altogether?

8   Jeanne swam $12\frac{1}{3}$ lengths of the pool and Keishla swam $14\frac{3}{4}$ lengths of the pool. How much further did Keishla swim?

9   What is the perimeter of a rectangular lawn measuring $12\frac{1}{4}$ m by $4\frac{4}{5}$ m?

10 Mr Butler is a keen gardener. He used $\frac{1}{8}$ of his garden for vegetables, $\frac{1}{2}$ for lawn and $\frac{1}{3}$ for fruit trees. What fraction of the garden is left for growing flowers?

Multiplying fractions

Explain

You already know that repeated addition can be done as multiplication.

When you multiply fractions, you multiply numerators by numerators and denominators by denominators. So:

$$\frac{1}{5} + \frac{1}{5} + \frac{1}{5} + \frac{1}{5} = \frac{4}{5}$$

$$4 \times \frac{1}{5} = \frac{4}{1} \times \frac{1}{5} = \frac{4}{5}$$

$$\frac{2}{5} + \frac{2}{5} + \frac{2}{5} = \frac{6}{5} = 1\frac{1}{5}$$

$$3 \times \frac{2}{5} = \frac{3}{1} \times \frac{2}{5} = \frac{6}{5} = 1\frac{1}{5}$$

Remember that any whole number can be written as a fraction with a denominator of 1.

Example 1

What is $\frac{3}{10}$ of 40?

Remember that the word 'of' means multiply in mathematics

$$\frac{3}{10} \times 40$$

$$= \frac{3}{10} \times \frac{40}{1}$$

$$= \frac{3 \times 40}{10 \times 1}$$

$$= \frac{120}{10} \quad \text{Multiply numerators, then denominators}$$

$$= 12 \quad \text{Simplify your answer}$$

Example 2

What is $\frac{3}{4} \times \frac{1}{2}$?

$$\frac{3}{4} \times \frac{1}{2}$$

$$= \frac{3 \times 1}{4 \times 2}$$

$$= \frac{3}{8}$$

The diagram shows that this is correct.

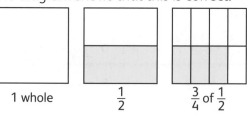

1 whole $\frac{1}{2}$ $\frac{3}{4}$ of $\frac{1}{2}$

This is $\frac{3}{8}$ of the whole.

If you **simplify** fractions before you multiply them, you can make the calculations easier, as you will be working with smaller numbers.

Look at these two methods of calculating.

Method 1: Simplify first

$$\frac{1}{3} \times \frac{9}{10}$$

$$= \frac{1}{3^1} \times \frac{9^3}{10} \quad \text{This is the same as } \frac{\div 3}{\div 3}$$

$$= \frac{3}{10}$$

Method 2: Simplify the answer

$$\frac{1}{3} \times \frac{9}{10}$$

$$= \frac{9}{30} \quad \text{Divide by } \frac{3}{3} \text{ to simplify}$$

$$= \frac{3}{10}$$

The way of showing the working in Method 1 is called cancelling.

Remember that to keep the fractions equivalent, you have to do the same to the numerator and denominator. So you can only cancel by dividing a numerator and denominator by the same number.

To multiply mixed numbers, regroup them and write them as improper fractions so that you can multiply numerators by numerators and denominators by denominators.

1 Calculate.

a $\frac{1}{2} \times \frac{3}{4}$ b $\frac{3}{4} \times \frac{5}{6}$ c $\frac{7}{8} \times \frac{7}{8}$ d $\frac{3}{5} \times \frac{2}{3}$

e $\frac{2}{3}$ of $\frac{1}{6}$ f $\frac{5}{6}$ of $\frac{3}{10}$ g $\frac{5}{8}$ of 16 h $\frac{5}{6}$ of $\frac{5}{8}$

2 Multiply. Simplify (cancel) where possible before you multiply.

a $\frac{8}{9} \times \frac{3}{5}$ b $\frac{7}{10} \times \frac{5}{8}$ c $\frac{1}{3} \times \frac{9}{20}$ d $\frac{1}{2} \times \frac{4}{5}$

e $\frac{1}{2} \times \frac{8}{10}$ f $\frac{3}{10} \times \frac{5}{6}$ g $\frac{5}{8} \times \frac{2}{10}$ h $\frac{2}{3} \times \frac{3}{4}$

i $3 \times \frac{5}{6}$ j $\frac{11}{12} \times 4$ k $\frac{3}{5} \times \frac{10}{13}$ l $\frac{7}{10} \times \frac{5}{21}$

3 Find the product.

a $\frac{3}{4} \times \frac{9}{10}$ b $\frac{15}{20} \times \frac{3}{4}$ c $\frac{18}{25} \times \frac{3}{4}$ d $\frac{3}{10} \times \frac{15}{16}$

e $\frac{49}{100} \times \frac{3}{7}$ f $\frac{3}{100} \times \frac{1}{3}$ g $\frac{1}{10} \times 10$ h $1\frac{1}{3} \times \frac{3}{4}$

4 Apply the correct order of operations rules to do these mixed calculations.

a $\frac{1}{2} \times \frac{3}{5} - \frac{1}{4}$ b $(\frac{3}{8} + \frac{1}{3}) \times \frac{3}{10}$

c $\frac{1}{4} \times \frac{2}{3} + \frac{2}{3}$ d $\frac{1}{3} \times \frac{1}{4} + \frac{3}{8}$

e $\frac{5}{9} - \frac{3}{10} \times \frac{5}{6}$ f $\frac{3}{5} + \frac{7}{10} - \frac{1}{2} \times \frac{4}{15}$

g $\frac{14}{15} \times \frac{5}{7} + \frac{4}{5}$ h $(\frac{2}{3} + \frac{5}{6}) \times \frac{1}{2}$ i $\frac{3}{4} - \frac{2}{5} \times \frac{1}{2}$

Think and talk

Do you remember the order of working when there is more than one operation? What operations do you perform first?

Problem-solving

5 There are 320 T-shirts in a delivery box. $\frac{1}{5}$ of the T-shirts are red, $\frac{1}{4}$ of the T-shirts are green and $\frac{3}{8}$ of the T-shirts are blue. The rest of the T-shirts are white.

a How many T-shirts are red?

b How many T-shirts are green?

c How many T-shirts are blue?

d What fraction of the T-shirts are white?

6 Josh has to walk $\frac{9}{10}$ of a kilometre to school. He is $\frac{1}{3}$ of the way there. How much further does he have to walk?

7 Mr Benson has 1 740 mangoes to sell. He has sold $\frac{5}{12}$ of them. How many has he sold?

8 $\frac{4}{5}$ of 320 students walk to school. How many do not walk?

9 What is the area of a rectangular flower bed that is $\frac{3}{5}$ m wide and $\frac{7}{8}$ m long?

10 Is it possible to enter fractions into a calculator? Investigate this and prepare short notes to summarise what you find out.

Dividing fractions

What is $\frac{1}{2} \div 5$? Look at the diagram carefully.

The pink section shows half of the shape.

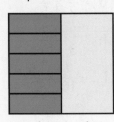 Each block in the pink section is $\frac{1}{5}$ of the half.

So, ▭ $= \frac{1}{10}$ of the whole.

$$\frac{1}{2} \div 5 = \frac{1}{10}$$

The product of a number and its **reciprocal** is always 1.

$$\frac{1}{5} \times \frac{5}{1} = \frac{5}{5} = 1 \qquad \frac{2}{5} \times \frac{5}{2} = \frac{10}{10} = 1$$

The reciprocal of a fraction is found by inverting the numerator and denominator.

So, if your fraction is $\frac{a}{b}$, its reciprocal is $\frac{b}{a}$.

We use reciprocals when we divide with fractions.

Dividing a fraction by another number is the same as multiplying the fraction by the reciprocal of the number you wanted to divide by. So we express the whole number as a fraction, then invert the fraction (switch the denominator and numerator) to get the reciprocal, and multiply.

This rule works when you divide a fraction by a whole number and when you divide a fraction by another fraction.

Example 1

$$\frac{5}{12} \div \frac{2}{3}$$

$$= \frac{5}{12} \times \frac{3}{2}$$

$$= \frac{15}{24} = \frac{5}{8}$$

Change the operator and invert the divisor

↓

Simplify

Example 2

$$\frac{7}{8} \div \frac{3}{4}$$

$$= \frac{7}{8} \times \frac{4}{3}$$

$$= \frac{28}{24} = \frac{7}{6}$$

$$= 1\frac{1}{6}$$

Change the operator and invert the divisor

↓

Simplify

To divide mixed numbers, first write them as equivalent improper fractions.

Example 3

$$2\frac{1}{2} \div 3$$

$$= \frac{5}{2} \div \frac{3}{1}$$

$$= \frac{5}{2} \times \frac{1}{3}$$

$$= \frac{5}{6}$$

Change the mixed number to an improper fraction

Write the whole number as a fraction

Example 4

$$3\frac{3}{4} \div 9 = \frac{15}{4} \div \frac{9}{1}$$

$$= \frac{15}{4} \times \frac{1}{9}$$

$$= \frac{15}{36}$$

$$= \frac{5}{12}$$

Change the mixed number to an improper fraction

Write the whole number as a fraction

---

1   Find the reciprocal of each number.

    a   3   b   11   c   5   d   144   e   18

    f $\dfrac{12}{13}$ g $\dfrac{2}{3}$ h $\dfrac{8}{9}$ i $\dfrac{5}{11}$ j $\dfrac{5}{18}$

2   Write each of these mixed numbers as equivalent improper fractions. Then write the reciprocal of each one.

    a $1\dfrac{11}{12}$ b $4\dfrac{3}{4}$ c $9\dfrac{9}{10}$ d $11\dfrac{10}{11}$ e $7\dfrac{1}{8}$

3   Express each of the calculations using a multiplication sign and work out the answers.

    a $\dfrac{3}{4} \div 2$ b $\dfrac{7}{8} \div 4$ c $\dfrac{9}{10} \div 3$ d $\dfrac{4}{5} \div 10$ e $\dfrac{9}{2} \div 4$

4   Find the value of n in each of these calculations.

    a $n \div 6 = \dfrac{1}{2}$ b $n \div 9 = \dfrac{1}{3}$

    c $n \div 8 = \dfrac{3}{4}$ d $n \div 12 = \dfrac{2}{3}$

> **Talk about it**
>
> Explain how you found the values of n in Question 4.

5   Calculate.

    a $\dfrac{15}{16} \div \dfrac{3}{4}$ b $\dfrac{1}{2} \div \dfrac{8}{9}$ c $\dfrac{14}{25} \div \dfrac{7}{15}$ d $\dfrac{9}{10} \div \dfrac{3}{8}$ e $\dfrac{17}{18} \div \dfrac{1}{3}$

    f $4\dfrac{1}{2} \div 6$ g $3\dfrac{1}{5} \div 5$ h $2\dfrac{9}{11} \div 3$ i $5\dfrac{1}{4} \div 7$ j $2\dfrac{1}{2} \div \dfrac{1}{4}$

Problem-solving

6   Jeanelle and Darvin train by running 6 km around a block in their neighbourhood. If the distance round the block is $\dfrac{6}{25}$ km, how many times do they run around the block to complete 6 km?

7   A teacher gives each student $\dfrac{2}{5}$ of a box of counters. If there were 6 boxes of counters, how many children were in the class?

8   Mr Garvey has a length of rope $25\dfrac{1}{2}$ m long. He cuts it into a number of length of $\dfrac{3}{4}$ m each. How many lengths can he make?

What did you learn?

Calculate.

1   $4 - 2\dfrac{3}{5}$ 2 $3\dfrac{3}{4} - 2\dfrac{1}{10}$ 3 $5\dfrac{1}{3} + 2\dfrac{6}{7}$

4   $\dfrac{2}{3}$ of 21   5 $\dfrac{3}{4} \times \dfrac{1}{5}$ 6 $\dfrac{12}{16} \times \dfrac{24}{30}$

7   $\dfrac{2}{3} + \dfrac{2}{3} \times \dfrac{7}{8}$ 8 $\dfrac{2}{3} \times (\dfrac{5}{8} - \dfrac{1}{4})$ 9 $\dfrac{5}{6} \times \dfrac{21}{25} + 1\dfrac{1}{5}$

10   $\dfrac{4}{3} \div 2$ 11 $\dfrac{9}{21} \div \dfrac{3}{7}$ 12 $4\dfrac{1}{4} \div 3$

B Calculating with decimals

Maths ideas

In this unit you will:
* add and subtract decimals using the column method
* multiply decimals and whole numbers
* divide decimals by whole numbers
* solve problems involving decimals.

Explain

To add or subtract decimals, make sure the **decimal points** are aligned below each other, then add or subtract as you would with whole numbers. Empty places may be filled with zeros.

It is useful to estimate by rounding before you calculate to check that your answer is reasonable.

Example 1

Calculate 12.045 + 2.3 + 0.8 + 6.

Estimate: 12 + 2 + 1 + 6 = 21

```
            Line up decimal points
  12.045
   2.300
   0.800  ← Write 0 in each empty place
 + 6.000  ← Remember 6 = 6.0
 ───────
  21.145
  11       Regroup  11/10 = 1.1
```

Example 2

Calculate 23.8 – 9.03.

Estimate: 24 – 9 = 15

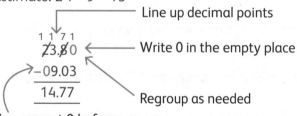

```
            Line up decimal points
   1 1 7 1
   23.80  ← Write 0 in the empty place
  –09.03
  ──────
   14.77     Regroup as needed
```

You can put 0 before whole numbers too if it helps

Think and talk

Jay says adding and subtracting decimals is easy as long you remember the PUP rule: Points Under Points.

a What does he mean when he says this?
b What do you do with empty spaces in columns when you add or subtract decimals?
c Can you place a 0 anywhere you like in a decimal fraction? Explain your answer.

1 Try to do these using mental methods. Write the answers only.

a 0.4 + 0.2	b 0.8 + 0.1	c 0.4 + 0.5	d 0.1 + 0.1
e 0.5 + 0.6	f 0.2 + 0.23	g 0.23 + 0.1	h 0.9 + 0.3
i 0.8 – 0.3	j 0.9 – 0.6	k 0.8 – 0.1	l 0.7 – 0.6
m 2 – 0.5	n 2.4 – 1.2	o 16.5 – 4.5	p 6.9 – 5.2

2   Write in columns and calculate.

    a   0.23 + 0.93   **b**   0.37 + 0.65   **c**   0.42 + 0.55   **d**   0.28 + 0.67

    e   1.49 + 0.99   **f**   2.34 + 0.07   **g**   4.09 + 2.8   **h**   5.32 + 5.48

    i   0.87 − 0.43   **j**   0.9 − 0.65   **k**   0.86 − 0.4   **l**   0.4 − 0.23

    m   14.3 − 3.09   **n**   12.09 − 4.5   **o**   5.45 − 0.99   **p**   3.67 − 2.8

    q   16 − 5.234   **r**   9 − 1.008   **s**   25 − 14.809   **t**   342 − 45.09

3   Estimate and then calculate.

    a   5.99 + 15.32 + 231.09   **b**   214.6 + 87.99 + 234.9

    c   612.5 + 132.09 + 99.5   **d**   231.9 + 54.3 + 9.085

    e   325.3 − 124.865   **f**   243 − 124.55

    g   412.89 − 128.805   **h**   100 − 45.087

4   Estimate and then calculate these decimal quantities. Remember to include units in your answers.

    a $5.87 + $500 + $235.50 + $100.99

    b   6.876 km + 500 km + 1 200 m (convert to km first!)

    c   56.004 seconds + 12.5 seconds

    d   78.13 kg − 32.95 kg

    e   300 litres − 234.565 litres

    f   12 cm + 123 mm + 3.4 cm + 0.9 cm (convert all units to cm first!)

Problem-solving

5   A shopkeeper buys 50 m of rope. He sells 13.5 m and 8.25 m.
    How much is left?

6   Percy is on a diet to improve his health. He loses 0.75 kg one week and 1.08 kg the next week.
    How much does he lose in total?

7   Debbie gets a bill for $28.75 and pays with two $20 bills. How much change will she receive?

8   A small plane flies 123.65 km on Friday, 132.08 km on Saturday and 109 km on Sunday.
    How far is this altogether?

9   Shamila has a mobile phone app that maps her route when she goes
    for a ride on her bicycle. The distance for each sector is shown. Look
    at the route map and work out how far Shamila cycled.

10 Passengers are allowed to carry hand baggage onto an airplane as
    long as it weighs less than 7 kg. Juan has a bag that weighs 1.2 kg.
    He packs a laptop of 2.45 kg and some books with a mass of 1.5 kg
    into the bag. His mom gives him a gift for his aunt of 0.459 kg and
    a packet of chips with a mass of 200 gram. He also has a bottle of
    water of 250 grams. Will his bag be allowed on board?

Multiplying and dividing decimals

Explain

You already know how to multiply and divide by powers of 10 mentally.

When you multiply by 10, each **digit** moves one place to the left to make the answer greater.

$15 \times 10 = 150$ $1.5 \times 10 = 15$ $1.05 \times 10 = 10.5$

Division is the inverse, or opposite, of multiplication. So when you divide by 10, the digits move one place to the right, making the answer smaller.

$150 \div 10 = 15$ $15 \div 10 = 1.5$ $1.05 \div 10 = 0.105$

When you multiply or divide by 100 or 1 000, you move digits one place for each power of ten. So for 100 you move two places and for 1 000 you move three places. The direction of the movement depends on whether you are multiplying or dividing.

To multiply decimals by other numbers, you follow three simple steps.

Step 1:	Step 2:	Step 3:
Count the total number of digits (**decimal places**) after the decimal point in each factor in the question.	Multiply the numbers as if there were no decimal points.	Place the decimal point in the product so that there are the same number of decimal places after the decimal point as there were altogether in the factors.

Example 1

5×3.43

$5 \times 3.\underline{43}$ There are two decimal places in the factors

$\begin{array}{r} {}^{2}\,{}^{1}\!343 \\ \times\ 5 \\ \hline 1715 \end{array}$

$5 \times 3.43 = 17.15$
↑
Insert the decimal point so the product has two decimal places

Example 2

0.7×0.8

$0.\underline{7} \times 0.\underline{8}$ There are two decimal places in the factors

$7 \times 8 = 56$ $0.7 \times 0.8 = 0.\underline{56}$
↑
Insert the decimal point so the product has two decimal places. Write 0 to show there are no whole numbers.

1   Try to do these mentally. Write the answers only.

a	3.6×10	b	6.5×10	c	0.4×10	d	1.2×10
e	$0.2 \div 10$	f	1.2×100	g	$456 \div 100$	h	0.08×100
i	$0.4 \div 10$	j	$235 \div 100$	k	$162 \div 100$	l	$250 \div 1\,000$

2   Calculate.

a	4.5×4	b	2.4×8	c	3.7×5	d	3.23×3
e	1.25×6	f	3.42×9	g	3.45×12	h	4.56×20
i	25.4×31	j	2.8×25	k	9.13×25	l	0.89×13
m	8.7×30	n	0.76×23	o	24×8.25	p	21×4.55

3   A builder needs 25 m of timber at $12.68 per metre. What will it cost him?

4   The overall cost of a wedding reception for 100 guests was $12 345. What was the cost per guest?

5   Ten thousand people each paid $1.99 for a charity raffle ticket.
   a   How much money was raised through ticket sales?
   b   The money from ticket sales was shared evenly among 100 different children's charities. How much did each receive?

Explain

When you divide a decimal by a whole number, place the decimal point in the quotient above the decimal point in the dividend. This works for both short and long division.

Example 1	Example 2
$126.8 \div 2$	$142.8 \div 12$

Example 1:
$$\begin{array}{r} 63.4 \\ 2\overline{)126.8} \end{array}$$

Example 2:
$$\begin{array}{r} 11.9 \\ 12\overline{)142.8} \\ -12\downarrow \\ \hline 22 \\ -12\downarrow \\ \hline 108 \\ -108 \end{array}$$

6   Calculate
   a $0.9 \div 3$ b $0.8 \div 4$ c $1.2 \div 4$ d $4.9 \div 7$
   e $6.4 \div 8$ f $0.36 \div 2$ g $3.6 \div 12$ h $0.24 \div 4$

7   Calculate. Make sure you have the correct number of decimal places in your answers.
   a $114 \div 5$ b $11.4 \div 5$ c $907.92 \div 12$ d $9\,079.2 \div 12$
   e $77.76 \div 32$ f $777.6 \div 32$ g $838.35 \div 15$ h $8\,383.5 \div 15$

8   What number am I? Work out the number for each of the following.
   a   If you divide me by 32 you get 3.65.   b   I am 5.63 less than 151.37 divided by 7.
   c   If you multiply me by 9, you get 10.35.   d   If you multiply me by 8, you get 11.76.

What did you learn?

1   Calculate.
   a $32.65 + $23.08 − $12.00   b   6.51 kg + 14 kg + 12.8 kg
   c $32.85 \div 15$ d 23×12.86

2   Rewrite these calculations with the decimal point in the correct position.
   a $12 \times 0.9 = 108$ b $1.9 \times 2 = 38$ c $0.8 \times 80 = 64$

3   In New York, the temperature at 6:00 a.m. was 12.8 °C. By noon, it was 4.8 degrees warmer and it increased by a further 2 degrees in the afternoon. By early evening, the temperature dropped by 2.9 degrees and it continued to decrease by another 8.4 degrees to midnight. What was the temperature at midnight?

C Mixed problems

Read the problems with your partner. Talk about how you would solve each one and then work out the answers in your book.

1   Mira carried three bags of shopping home from the supermarket for her gran. The bags weighed $2\frac{1}{10}$ kg, $3\frac{3}{4}$ kg and $2\frac{4}{5}$ kg. What was the total mass of the bags?

2   $1\frac{4}{5}$ litres of water is poured from a $2\frac{1}{2}$ litre container. How much water is left?

3   It takes Earth one year, or 364.25 days, to complete one revolution around the Sun. It takes Venus 0.62 Earth years to complete one revolution.

   a   How many days will it take for Earth to complete five revolutions?

   b   How many days does it take Venus to complete one revolution?

4   Each piece of wood must be cut into the number of equal pieces shown below it. Work out the length of the pieces. Give your answers in both metres and centimetres.

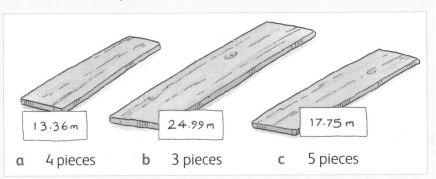

    13.36 m   24.99 m   17.75 m

   a   4 pieces   b   3 pieces   c   5 pieces

5   Each roll of cloth needs to be divided equally into the number of equal pieces shown below it.

    21.96 m   49.20 m   35.88 m

   i   12 pieces   ii   15 pieces   iii   13 pieces

   a   Work out the length of each piece.

   b   Work out how much money the seller will make if she sells 15 pieces of fabric for $23.65 each.

What did you learn?

1   Make up three problems involving adding and subtracting fractions.

2   Make up three problems involving multiplying and dividing decimals.

3   Swap your problems with a partner and solve each other's problems.

Topic 6 Review

Key ideas and concepts

Complete these statements in your own words to summarise what you learnt in this topic.

1 When you add or subtract fractions it is important to _____.

2 When you multiply a fraction by another fraction you must remember to _____.

3 Mixed numbers are easier to work with if you _____.

4 A reciprocal is _____.

5 You don't ever really have to divide fractions because _____.

6 To add or subtract decimals you _____.

7 The most important rule for multiplying decimals is _____.

8 You can divide decimals just like whole numbers as long as _____.

Think, talk, write …

1 Answer these questions in your maths journal.
   a What did you find easiest in this topic? Why?   b What did you find most challenging? Why?

2 Discuss these questions in groups.
   a What three things are important when you work with fractions? Why?
   b How can you use estimation to decide whether your answer to a calculation involving decimals is reasonable or not?
   c Is it true that you can just ignore decimal points when you are multiplying decimals? Explain your answer.

Quick check

1 Do these calculations as quickly and efficiently as you can.

   a $\frac{4}{5}+\frac{1}{3}$ b $2\frac{2}{5}+\frac{3}{10}$ c $\frac{3}{4}-\frac{3}{8}$ d $2\frac{3}{4}-1\frac{7}{8}$

   e $\frac{2}{3}$ of 60   f $\frac{3}{4}$ of \$24   g $\frac{3}{4}$ of \$10   h $\frac{7}{8}\times\frac{3}{5}$

   i $\frac{2}{3}\times\frac{5}{8}$ j $\frac{5}{3}-\frac{3}{4}+\frac{1}{6}$ k $\frac{5}{3}-(\frac{3}{4}+\frac{1}{6})$ l $\frac{2}{3}+\frac{1}{3}\times\frac{1}{4}$

   m $\frac{4}{5}\div6$ n $2\frac{1}{3}\div4$ o $2\frac{1}{2}\div\frac{1}{3}$

2 Calculate.
   a $4.7+12.65+0.812+12$ b $18.8-9.25$ c $2-0.875$
   d 1.5×6 e 0.12×7 f 0.2×12
   g $2.46\div3$ h $0.27\div3$ i $1.75\div13$

Problem-solving

3 What is the cost of 13 litres of cooking oil at \$5.50 per litre?
4 A rocket travels at 88.5 m per second. How far will it travel in 25 seconds?

Test yourself (1)

Explain

Complete this test to check that you have understood and can manage the work covered in Topics 1 to 6.
Revise any sections that you find difficult.

1   Write these as numerals.
   a   Three million twenty-five thousand two hundred and three
   b   Ninety-nine thousand four hundred and sixty-three

2   Write each number in words.
   a   2 312 065   b   4 100 876   c   9 098 089

3   Measure each angle in degrees and state what type of angle it is.
   a   b   c

4   Use rounding to estimate the answers, then calculate.
   a   603 + 715 + 986
   b   7 899 − 5 211
   c   24 999 ÷ 2
   d   408 × 31

5   Simplify each fraction.
   a $\dfrac{33}{3}$ b $\dfrac{11}{2}$ c $\dfrac{25}{60}$ d $\dfrac{8}{100}$

6   Calculate and give the answers in simplest form.
   a $1\frac{1}{2} + 2\frac{2}{3}$ b $6 - 4\frac{2}{3}$ c $\frac{4}{5} \times \frac{2}{6}$ d $2\frac{3}{4} \div 4$

7   Change each decimal to a fraction or mixed number in simplest form.
   a   0.3   b   0.45   c   4.25   d   5.44

8   Calculate.
   a   1.9 + 2.43   b   18 − 2.75   c   0.2 × 0.02
   d   1.84 × 10   e   23.45 × 100   f   7.24 ÷ 4
   g   245 ÷ 100   h   3.8 ÷ 0.2   i   0.75 of 420

9   Round these numbers as instructed.
   a   0.786 to the nearest tenth
   b   194.123 to the nearest hundredth
   c   19.15567 to two decimal places
   d   0.999 to two decimal places

10 Look at the shape below.
    a What is the mathematical name for this shape?
    b How many vertices does this shape have?
    c How many angles does it have?
    d How many pairs of parallel sides does it have?

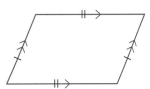

11 Write each percentage as a fraction or mixed number in simplest form.
    a 70%   b 17%   c 140%   d 2.5%

12 Write as a percentage.
    a $\dfrac{1}{2}$ b $\dfrac{4}{5}$ c $1\dfrac{1}{4}$ d $\dfrac{48}{200}$

13 Write each percentage as a decimal fraction.
    a 85%   b 5%   c 25.5%   d 200%

14 For each of the following shopping lists:
    a estimate the total cost
    b estimate how much change you would get from $100.00
    c work out the exact answers.

List A	List B	List C	List D
1 × $4.10	$56.09	3 × $3.79	6 × $2.47
4 × $0.97	2 × $2.20	2 × $1.49	$2.75
2 × $3.15	$2.76	$1.98	$4.70
1 × $6.59	$1.95	4 × $1.54	4 × $0.99
	10 × $1.06		$3.80

15 Say what type of triangle each of these is and give reasons for your choice.

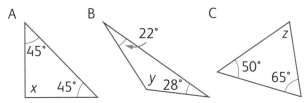

16 Calculate.
    a 32 876 ÷ 23   b 32 456 × 19
    c 124 567 + 45 987   d 342 098 − 12 987

17 A lion can run at $\dfrac{3}{5}$ of the speed of a cheetah. If a cheetah's speed is 115 km/h, how fast can the lion run?

18 A bar of gold with a mass of 16.68 kg is divided into 8 equal pieces. What is the mass of each smaller piece?

19 Mrs Jones buys 300 m of wood at $40.78 per metre. What is the total cost?

20 Linda paid $156 for 400 plastic containers. How much was each container?

21 A person is paid $20.67 per hour. If she works 38 hours, how much will she earn?

73

Teaching notes

The metric system

* The units in the metric system work in powers of 10 and allow you to write measurements in equivalent ways.

* Students need to realise that when they convert to a larger unit (for example, from grams to kilograms) they will have a smaller number of that unit and when they convert to a smaller unit they will have a larger number of those units.

Imperial and customary units

* The imperial system of measurement dates back to the 1800s in England. The US customary units developed from that, and while some units are the same, the US system has different units for dry good and liquids. An imperial gallon, for example, is about 4.5 litres, but a US gallon is 3.8 litres.

* Students need a basic understanding of these units and how they are used informally, but they are expected to measure in metric units.

Estimating and measuring (length, mass and capacity)

* Students need ongoing and regular practice to estimate and measure accurately. Help students improve their skills by working with concrete objects and real-life situations. Use measuring instruments marked in different units to encourage converting between these.

* Explain to students that we often talk about the weight of something when we really mean its mass. While we may say 'I weigh 40 kilograms', mass and weight are technically not the same.

* Remind them that the capacity of a container is a measure of how much it can hold in total. When a container is partly filled, we talk about the volume of liquid it contains.

Scale drawings

* Scale drawings are representations of the real world in which everything is drawn at the same fraction of its actual size. All parts of the smaller object will have been scaled down by the same amount, so the proportions remain similar.

A

Look at the photos carefully. Why does the same item show two different masses on the scale? Which unit shown here is a metric unit? What type of unit is the other one? Use the two masses to try and work out how many grams there are in one ounce.

B

How do you know that this is not a photograph of a real car? What is the length of the car in the photograph? What do we call it when we make small models or diagrams of real objects? A real Mini is 390 cm long. How much smaller is this model?

C

The average (mean) mass of a single banana is 180 grams. Estimate the mass of this bunch in kilograms. Tell your partner how you got to your answer.

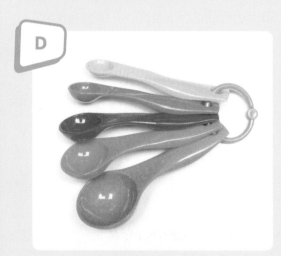

D

What are these measuring instruments called and what are they used for? Which of these holds the most? How much does it hold? Which holds the least? How much does it hold?

Think, talk and write

A   **Measuring systems and units** *(pages 76–80)*

1   Which of the following are metric units? What are the others?

gallons	litres	feet	grams

2   If you were building a new road, why would you need to take accurate measurements? What would you need to measure?

3   Which units of measurement would you use to make accurate measurements of the following:
   a   the mass (weight) of a suitcase full of clothes
   b   the amount of liquid in a medicine bottle
   c   your body temperature
   d   the height of a wall around your house.

B   **Length and scale diagrams** *(pages 81–84)*

1   What do you measure if you need to know your height? What unit or units of measurement would you use to do this?

2   What does the word distance mean?

3   Which stick is longer: a stick that is 1.2 m long or a stick that is 12 cm long?

4   What would you need to do to draw a scaled diagram of the classroom? Discuss this in groups.

C   **Mass** *(pages 85–86)*

The mass of trucks and other large vehicles is often measured on a device called a weighbridge. How do you think this works? What units of mass do you think it uses? Do some research to find out, if you do not know the answers.

D   **Capacity** *(pages 87–88)*

1   Can you measure the capacity of a triangle? Explain your answer.

2   A sign in an elevator says: Maximum capacity 8 persons or 650 kilograms. What do you think this means?

A Measuring systems and units

Explain

The table summarises some of the different units of measurement that you have previously worked with.

Measurement	Units
Length or distance	millimetre (mm), centimetre (cm), metre (m), kilometre (km)
Capacity	millilitre (mℓ), litre (ℓ), kilolitre (kℓ)
Mass	milligram (mg), gram (g), kilogram (kg), tonne (t)
Temperature	degrees Celsius (°C), degrees Fahrenheit (°F)

Maths ideas

In this unit you will:
* revisit the metric system and units of measurement
* convert between units in the metric system
* understand how metric units are related to other units of measurement.

Key words

metric system

customary units

imperial units

convert

1 Work in groups. Your teacher will give you a measurement topic. Discuss and list five different situations in which you would use this measurement. State the unit of measurement you would use in each situation.

2 Write down three things that you could weigh in each of these units:

    a milligrams   b grams   c kilograms.

3 Choose the unit that is the best choice to measure each of these items.

> litres (ℓ), millilitres (mℓ) or kilolitres (kℓ)

    a The amount of cola in a can   b The amount of water in a fish tank
    c The amount of gas needed to fill the tank of a truck   d The amount of crude oil in a tanker ship

4 Put these items in order from the smallest to the largest mass and distance.

    a Mass

        i A small car   ii A bag of sweet potatoes   iii A laptop computer

    b Distance

        i The distance a car can drive in 5 minutes

        ii The distance from the halfway line on a football field to the goalposts

        iii The distance a person can walk in 10 minutes

5 Work in pairs or groups. Your teacher will ask you to make some real measurements of length or distance.

    Decide which unit of measurement is best for each item or distance.

    * Estimate each measurement and then take accurate measurements.

    * Record your results in a table like this one.

What am I measuring?	Unit of measurement	Estimate	Accurate measurement

Work in groups to discuss and solve these problems.

6   You have to send some documents by post.

The Post Office charges by the size and weight of each item you post. If your letter is 220 mm by 320 mm and weighs 500 grams, use the information on the right to work out how much you will pay.

Postage costs: Priority mail		
Size	**Weight**	
220 × 110 mm	first 100 g	$ 1.60
	each additional 50 g	$ 0.70
229 × 324 mm	first 100 g	$ 2.20
	each additional 50 g	$ 0.80

7   Imagine that you are farmers or traders and that there is no common unit of measurement for mass. You want to trade the goods that you grow or produce. What could you use as a unit of measurement? What would you need to think about?

8   You have a factory that produces potato chips. You want to change the packaging of the chips. To do this you need to work out how many chips (more or less) could fit in each bag. How would you do this?

9   Your school is preparing for a carnival. Students from several other schools will be coming to your school to join in the celebrations. You need to work out how many students can stand on one of the school sports fields at the same time. The students need to be able to swing their arms without hitting each other. How could you work this out?

Explain

Different systems of measurement

The **metric system** is the official system of measurement in most countries of the world. It is a decimal system, and units increase or decrease in size by powers of 10.

In the metric system, the kilogram is the standard unit of mass. This is a metric measure that was first used in France in 1799. At that time, it was decided that one kilogram should be equal to the mass of one litre of water at 4 °C. Later, scientists decided to make a kilogram in metal to use as a standard. This kilogram mass is kept under three bell jars in a specially guarded place in France.

There are only three countries in the world that have not officially adopted the metric system: the USA, Liberia and Myanmar. In the USA, the measurements are called **customary units**, elsewhere they are called **imperial units**.

Customary units are not metric. They include units such as gallons, quarts, pints, inches and feet.
*   1 quart (qt) = 2 pints (pt)
*   1 gallon (gal) = 4 quarts (qt)
*   1 foot (ft) = 12 inches (in)
*   3 feet = 1 yard (yd)

Customary units such as feet, inches, yards and miles are used to measure length or distance. Pounds and ounces are used for mass and pints and gallons for capacity.

Although your country officially uses the metric system, you may find people still using imperial units of measurement, so it is important to understand what these mean.

Investigate

10 Work in pairs. Find the answers to the questions below. Then compare your answers.
   a   Where did the imperial system of measurement come from?
   b   Why is the metric system used in so many countries?
   c   If a person weighs 8 stone, how much do they weigh in kilograms? (You can find many apps and programmes to convert measurements on the Internet.)

11 Work in pairs. Which person is the tallest? You will need to convert feet and inches to metres in order to work this out.

12 Look at this graph carefully.

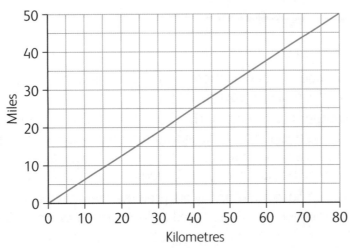

   a   Explain what this graph shows.
   b   Use the graph to convert 40 kilometres to miles.
   c   How many kilometres are 25 miles?
   d   Which is further, 60 km or 45 miles?

Explain

Converting units of measurement

Metric units: To **convert** metric units, you divide or multiply by 10 or powers of 10 (such as 100 or 1 000).

* To convert millilitres to centilitres, you multiply by 10.
* To convert milligrams to grams, you multiply by 1 000.
* To convert metres to centimetres, you divide by 100.

Decimal notation: You can convert units in decimal notation by moving the digits left or right. Use what you already know about place value. The numerals stay the same.

> **Examples**
> 1.000 km = 1 000 m
> 1.700 km = 1 700 m (1.7 × 1 000)
> 2 300 ml = 2.300 l (2 300 ÷ 1 000)
> 4 000 g = 4.000 kg (4 000 ÷ 1 000)

13 Study the step diagram. Use it to convert these units.

a   25 decilitres = _____ litres

b   3 kilograms = _____ grams

c   55 millimetres = _____ centilitres

d   250 centigrams = _____ grams

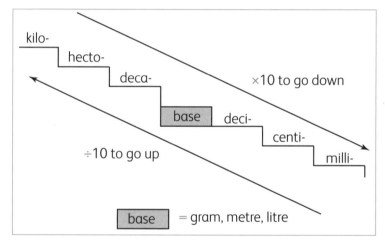

14 Work in pairs. Study the diagram, which shows how to convert metric units for distance and length. Explain to your partner how this works. Then make up questions to ask each other. For example: An animal is 2.7 metres long. How many centimetres is that?

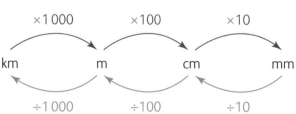

15 Convert these measurements.

a   35 kg = _____ g

b   560 g = _____ kg

c   _____ g = 0.450 kg

d   25 g = _____ kg

e   _____ ml = 25 l

f   3 200 ml = _____ l

g   0.375 l = _____ ml

h   1.5 l = _____ ml

i   1.5 km = _____ m

j   2 500 m = _____ km

k   35 mm = _____ cm

l   _____ cm = 3.2 m

16 In a science lesson, you follow instructions to make two solutions. You put 1 500 milligrams of salt in one 100 mℓ jug of water and 150 grams of salt in another 100 mℓ jug of water. Which solution has more salt?

17 Joni bought some pieces of ribbon in a sale. Put the lengths of ribbon in order from the shortest piece to the longest piece.

| 1.15 m | 350 cm | 2 500 mm | 1.20 m | 0.5 m |

18 Work in pairs and discuss how to complete these statements.
   a To convert pints to quarts, _____ by _____.
   b To convert inches to feet, _____ by _____.
   c To convert quarts to gallons, _____ by _____.

19 Convert these measurements.
   a 5 foot 3 inches = _____ inches
   b 6 pints = _____ gallons
   c 8 quarts = _____ pints
   d 24 inches = _____ feet
   e 2 yards = _____ inches

20 Work in groups. Create 10 word problems involving measurement. Make sure you know the correct answers. Then have a quiz where you test another group using the questions you have created. Discuss the rules of the quiz before you begin, including how you are going to score the answers.

What did you learn?

1 Here are some units of measurement:

| pound | gallon | kilogram | gram | stone | ounce | milligram |

   a Which of these units could you use to measure mass?
   b Which units are metric and which are customary units?
   c Which units would you use to measure a large mass?

2 Convert each of these metric measures to the base unit of that measurement.
   a 4 000 mℓ
   b 23 000 cm
   c 987 000 mg
   d 24.5 km
   e 19.4 kℓ
   f 235 hectograms

3 If you have two gallons of water and your friend has two litres of water, who has more water? What is the difference?

4 Write a paragraph to explain the difference between metric units and customary units.

B Length and scale diagrams

Explain

You already know that **length** is a measure of how long, wide, tall or far something is.

Length, width, height and **distance** are all types of length.

You can measure short lengths in centimetres (cm) or millimetres (mm).

Longer lengths and distances are usually measured in metres or kilometres.

Maths ideas

In this unit you will:
* estimate and measure lengths and distances
* convert and work with units of length
* revisit scale diagrams and use them to work out actual distances.

1 What is the most appropriate unit for measuring these lengths?

a

Width of a finger

b
Height of a person

c
Length of an ant

Key words

length
distance
scale diagram
maps
plans
scale

d

Distance from one island to another

e

Length of a roll of fabric

f
Distance from one town to another town

Remember

1 cm = 10 mm
$\frac{1}{2}$ cm = 0.5 cm = 5 mm
1 metre = 100 centimetres
1 kilometre = 1 000 metres

g
Thickness of an eyelash

h
Length of a foot

i
Height of a building

2 Accurately measure each of these lengths, first in centimetres and then in millimetres.

a

b

c

d

3 Estimate each of these lengths and write down your estimates. Compare your estimates with a partner's. How can you check whose estimates are more accurate?

a The length of your longest finger
b The height of your classroom
c The distance from school to the nearest airport
d The height of the tallest tree in your street

4   Liz bought 3.84 metres of ribbon on Monday and 676 cm of ribbon on Tuesday. How many metres of ribbon did she buy altogether?

5   John walked 2.67 km. Henry walked 3 125 m.
   a   Who walked further?
   b   How many metres further did he walk?

Investigate

6   Work with a partner to plan the route for a fun run that is 500 metres long at your school.

   *   Use measuring tapes or metre sticks to measure out the route.

   *   Draw a simple map of the route.

   *   How long do you think it will take a Level 6 student to run this distance? What things could affect your answer to this question?

Explain

A **scale diagram** is an accurate drawing that shows something at a smaller size than it really is. **Maps** and **plans** are scale diagrams of the real world.

When you draw a map, you have to draw things smaller than they really are to fit them onto a page. The distances on a map are a fraction of the real distances. The map **scale** tells you how much smaller the map is than the distance in the real world.

The scale of the map $= \dfrac{\text{length on the map}}{\text{length in the real world}}$

The scale of a map can be given in words, for example, 1 cm = 5 km.

This means that 1 cm on the map represents 5 km in the real world.

Scale can also be expressed as a ratio such as 1 : 1 000, which means that 1 unit on the map represents 1 000 of the same units in the real world. So, 1 cm on the map represents 1 000 cm in the real world and 1 mm on the map represents 1 000 mm in the real world.

On many maps, you will find a scale bar like this one:

0	100	200	300	400	kilometres
0		100		200		300 miles

The small divisions on the scale bar show map distances, but the number tells you what the real distances are. You can compare distances on the map with the scale bar to work out how far they are in reality.

7   This is a scale diagram comparing the height of three famous structures.

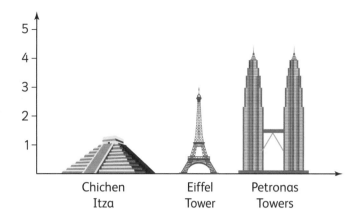

a   Why are diagrams like this called scale diagrams?

b   If one division on the diagram scale represents 100 m in the real world, estimate the real height of each structure.

c   How many times larger are the heights in the real world?

8   Express these scales in words.

a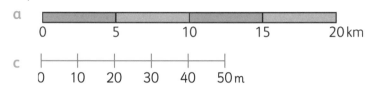

b   1 : 25 000

c

9   Measure these lines in centimetres. Write your answers in decimal form.

a   _____

b   _____

c   _____

d   _____

10 In the lines above, 1 cm represents 500 metres. What distance does each line show?

11 On this map, 1 cm represents 100 metres. Use your ruler to measure the distances. Give your answers in metres.

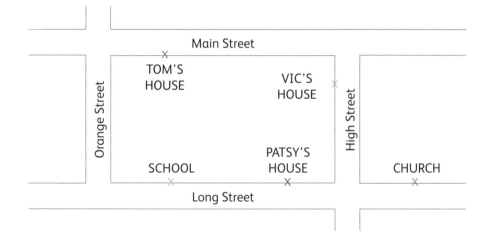

How far is it from:

a   Tom's house to school?

b   Vic's house to school?

c   Patsy's house to school?

d   Tom's house to Vic's house?

e   Vic's house to church?

f   Patsy's house to church?

g   Patsy's house to Vic's house?

h   Tom's house to church?

12 Use the map from Question 11 again.

a   Find the shortest route from Tom's house to Patsy's house.

b   How far is this in metres?

13 This map shows the roads in a small town.

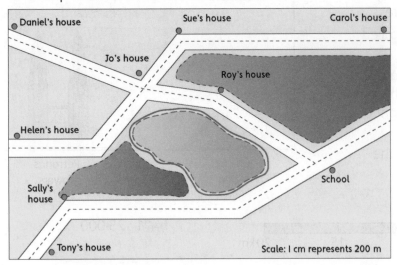

Work out these distances in metres. Measure along the roads.

a   From Carol's house to Tony's house

b   From Sally's house to school

c   From Sally's house to Carol's house

d   From Daniel's house to Tony's house

e   From Roy's house to Tony's house

f   From Jo's house to school

g   From Helen's house to Daniel's house

14 Sue and Carol walk across the field to get to school.

a   How many metres does Sue walk?

b   How many metres does Carol walk?

c   If they walked along the road, how much longer would their route be?

15 On a map, 1 cm represents 5 km. If the actual distance between two towns is 40 km, what is the length on the map?

16 The actual distance between two islands is 96 km. If 1 cm on a map represents 6 km, how far apart are the towns on the map?

17 The actual distance between two towns is 55 km. If the map scale is 1 : 1 000 000, what is the distance between them on the map?

18 On a house plan, 1 mm represents 40 cm in reality. If the length of the house is 30 metres, what is the length of the house on the plan?

What did you learn?

1   Measure the lengths of the sides of your desk to the nearest centimetre. Use a scale of 1 cm to show 10 cm to draw a scale drawing of your desk.

2   Use a scale of 1 cm to 60 cm. Draw lines to represent:

   a   60 cm   b   360 cm   c   150 cm.

3   The distance between two towns on a map is 65 mm. If the map scale is 1 : 25 000, what is the real distance between the two towns?

C Mass

A common unit of measurement for **mass** is the **gram** (g).
A large paperclip weighs about 1 gram (g). Some things are
even lighter than this. We measure them in **milligrams** (mg).
We measure heavier things in **kilograms** (kg) and **tonnes** (t).

Maths ideas

In this unit you will:
* estimate and measure mass in kilograms, grams and milligrams
* compare the mass of different objects using different units of measurements.

1   Choose the most suitable mass from the box for each item shown.

| 300 g   7 kg   1 tonne   200 g   3.5 kg   1 gram   10 tonnes   1 kg |

Key words

mass	kilograms
gram	tonnes
milligrams	

a    b    c   d

e    f   g    h

Remember

1 t = 1 000 kg
1 kg = 1 000 g
1 g = 1 000 mg

2   Judy went shopping. She brought home three bags of groceries.
Calculate the total mass of the contents of each bag. Give your
answers in kilograms.

a   550 g   +   1.2 kg   +   233 g

b   420 g   +   1.5 kg   +   908 g

c   960 g   +   850 g   +   500 g

3   Convert each mass to the units shown.
    a   420 g to kg   b   1 518 mg to g   c   1.22 t to kg
    d   37 kg to t   e   0.4 g to mg   f   0.08 kg to g

4   A pumpkin weighs 2.4 kg. It is cut into five pieces of equal mass. Find the mass
    of each piece. Give your answer in grams.

5   A truck weighs 3.18 tonnes. When carrying its load, the truck weighs 4.72 tonnes. Find the mass of the load in tonnes.

6   A potato weighs 400 grams. There are 42 potatoes in a box. If the empty box weighs 0.9 kg, what is the mass of the box of potatoes? Give your answer in kilograms.

7   A bag holds 56 kg of rice. A packet of rice holds 500 grams. How many packets of rice can be filled from the bag?

8   Mrs Jones has a mass of 85 kilograms. Her doctor advises her that her ideal mass should be 70 kg.

   a   How many kilograms should Mrs Jones lose to reach her ideal mass?

   b   If she aims to lose the total amount over 12 weeks, how many kilograms should she aim to lose each week?

9   The mass of a small brick is about 625 g. A builder orders 2 tonnes of bricks. Approximately how many bricks does he get?

Explain

Do you remember how to read mass on a scale marked in different intervals?

This scale is marked in kilograms.

The shorter lines between the kilogram markings represent half kilograms.

The needle on this dial is halfway to the half kilogram mark, so it is showing a mass of 1.25 kg (remember that $\frac{1}{4}$ = 0.25).

10   Write the masses as shown on each scale.

   a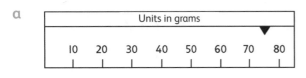

   b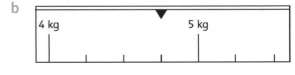

   c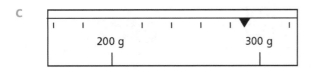

   d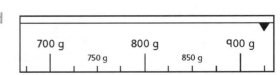

What did you learn?

1   Write these masses in order from lightest to heaviest.

   | 2.6 kg | 2 500 g | 4 000 mg | 800 g | $2\frac{7}{8}$ kg |

2   30 identical toys have a mass of 3.6 kg. What is the mass of each toy?

3   A cricket ball has a mass of 160 g and a tennis ball has a mass of 58 g. How many of each type of ball would you need to make 1 kilogram?

D Capacity

Explain

The amount of liquid that a container can hold is called its **capacity**. We can use **litres** (ℓ), **centilitres** (cℓ) and smaller units called **millilitres** (mℓ) to measure how much liquid a container holds.

1   List four objects whose capacity can be measured in litres.

2   Would you measure the capacity of each of these containers in litres or millilitres?

a

b

c
MILK

d

e

3   Write these in litres, using decimals.

a	125 mℓ	b	15 mℓ	c	2.5 mℓ	d	306 mℓ
e	2 498 mℓ	f	51.39 mℓ	g	31 000 mℓ	h	216.8 mℓ

4   Convert to millilitres.

a	1.960 ℓ	b	3.006 ℓ	c	16 ℓ	d	2.4 ℓ
e	0.67 ℓ	f	81 ℓ	g	0.057 ℓ	h	0.3 ℓ

5   Write the amount of liquid in each jug in mℓ.

A

B

C

D

Problem-solving

6   An urn has a capacity of 3.3 litres. How many 200-mℓ cups of water can you pour from a full urn?

7   Mrs Smith takes 5 mℓ of medicine four times a day. The bottle she has contains 140 mℓ of medicine. How many days will the bottle last?

8   Shayne has the following amounts of liquid left in four containers. He wants to pour them into a drum that holds 15 litres.

15 ℓ   5 370 mℓ   3.5 ℓ $1\frac{3}{4}$ ℓ   3 708 mℓ

a   Calculate the total amount of liquid in the four containers in litres.

b   How much more liquid will he need to fill the 15-litre drum?

9   A hotel swimming pool has a capacity of 900 kilolitres. About 200 litres of water are lost through splashing and evaporation each day.

a   How much water is lost from the pool in a week?

b   The pool is full on Monday and it is only topped up three days later. How much water is left in the pool just before it is topped up?

10   Mike's car uses 1 litre of petrol per 12 km (on average).

a   How many litres will the car use if he travels 200 km?

b   The petrol tank has a capacity of 45 litres. How much petrol will Mike need to fill the tank if he's travelled 400 km since the last time he filled up?

What did you learn?

1   Write down three examples of containers that have their capacity given in millilitres.

2   How many millilitres are there in 5 litres?

3   Sally bought 18 containers of juice, each containing 330 mℓ. How many litres of juice is this in total?

Topic 7 Review

Key ideas and concepts

Answer these questions to summarise what you learnt in this topic.

1   What metric units can you use to measure short lengths?

2   What is the relationship between feet, inches and centimetres?

3   What would you measure in milligrams?

4   How many milligrams are there in $2\frac{3}{4}$ grams?

5   What does it mean if a container has a capacity of a kilolitre?

6   What does a scale of 1 : 200 mean on a diagram?

Think, talk, write ...

1   Discuss how each of these people might use scale diagrams in their work.

architect   town planner   ship's captain   dress designer

2   Write short notes to teach someone how to convert between measurements in the metric system.

3   Do people in your community use customary or imperial measures? Give examples.

Quick check

1   Which units of measurement would you use to measure the following?
   a   The length of a road
   b   The capacity of a glass

2   Copy and complete this table of metric units of length.

	Hecto-	Deca-	Metre			
1 000 m		10 m		0.1 m		0.001 m

3   Rewrite each set of measurements in ascending order.
   a   32 cm   0.025 cm   1.8 cm   330 mm   0.3 m
   b   150 cm   2.2 m   37.8 cm   1 300 mm   0.99 m

4   The wheel of Micah's bicycle travels 1.35 m each time it turns. Work out how far he cycles if the wheel turns:
   a   100 times   b   1 000 times.

5   A bucket holds 3.4 ℓ and a full beaker of water holds 100 mℓ. How many beakers of water would you need to fill the bucket?

6   The scale on a map is 1 : 20 000. What distance on the map will represent 2 km?

7   A map uses a scale of 1 cm = 20 miles. Two towns are 6.5 cm from each other on the map. What is the actual distance between the two towns in miles?

8   You want to add a location to a map with a scale of 1 cm = 15 km. The location is 45 km from the city. How many centimetres away from the city will you mark the location?

8 Data handling (1)

Teaching notes

Methods of collecting data

* Students already know how to collect data by doing observations, drawing up a questionnaire and conducting an interview.

* Observation involves looking and often counting. A questionnaire is a form that is used to collect and record data. An interview involves asking questions to collect data.

* Students need to think carefully about what they are trying to find out to choose the most appropriate method for collecting the data. In some cases, they may need to use other sources (such as books, websites or television programmes) to find the data they need. Data collected from other sources is called secondary data.

Organising data

* Sets of numbers are not really useful or easy to interpret. Data is more useful if it is sorted and organised. Students already know how to use tallies (////) to record data items in a table or chart.

* A frequency table is a way of organising data to show how many (the frequency) data items there are in each category or group.

Averages

* Averages give you a general idea of what a data set tells you, for example, the average salary of office workers, or the average time a student spends watching TV each night.

* There are different types of averages. The mean is an arithmetic average: to find the mean you work out the sum of the data values and then divide the answer by the number of data items. The mode is the item that occurs most often in a data set and the median is the middle value when data is arranged in size order.

A

Mr James owns an ice-cream shop. He wants to reduce the number of flavours he sells to reduce wastage. How could he use data to decide which flavours to keep and which flavours to stop selling? What methods could he use to find the data he needs to make these decisions?

You read in Topic 7 that the average mass of a banana is 180 grams. What does this mean? How do people work out an average like this? If you weighed a banana and found its mass to be 230 grams, what could you say about it?

Think, talk and write

A Collecting and organising data
(pages 92–94)

1   Shania collected data to see how many students had a birthday in each month. These are her results:

January	March	August
November	December	June
August	June	June
February	June	August
January	November	September
August	February	July
April	May	November
September	August	January
October	November	October
July	April	April
May	December	

a   What methods could she have used to collect the data?

b   How many people did she collect data from?

c   Why is this data set not so easy to work with?

d   What could you do to make the data set easier to work with?

e   Draw a table to summarise this data set and make it easier to work with.

2   Brainstorm a list of sources of information that you could use to find out more about imports and exports in your country.

B Averages *(pages 95–96)*

1   What does the word 'average' mean in these statements?

  a   The average score for the test was 22.

  b   The average temperature in December was 25 °C.

  c   Young children in this town spend an average of 3.5 hours a day watching television.

2   If you get the following marks for three maths tests, what is your mean test mark?

  $\frac{15}{20}$ $\frac{14}{20}$ $\frac{17}{20}$

A Collecting and organising data

Explain

You can **collect data** or information from many sources, for example, from **observations** or **interviews**, or by doing research and using published or publicly available information (books, websites, sales figures and so on).

If you want to collect data, you can conduct a **survey**. You will need to create a **questionnaire** and record and organise the results. You can do this using a **tally chart** and a **frequency table**. You can then draw a **graph** and analyse the data you have collected.

Maths ideas

In this unit you will:
* revisit different methods of collecting data and decide what type of data suits each method
* collect your own data using observation, interviews and questionnaires
* use tallies and frequency tables to record and organise data.

Key words

collect
data
observations
interviews
survey
questionnaire
tally chart
frequency table
graph

1   Read the questions and decide which question is better for each survey. Say why the other question is not suitable.

   a   A survey to find out what food is the most popular in the school canteen:

      *   Do you buy food from the school canteen?

      *   What food do you buy at the canteen?

   b   A survey to find out about favourite colours:

      *   What colour do you like?

      *   Do you like pink or blue best?

2   For each set of data below, decide which sources of information and which method is most suitable for collecting the data.

| The sports games that will be on television during the next week | The land area of my country and the five nearest Caribbean countries |

| Favourite sports teams of the children at my school | Interest rates local banks pay on saving accounts |

| How children in my school get to school (for example, by walking, by bus or by taxi) |

Possible sources	Some methods
Reference books	Reading
The Internet	Observation
The radio	Asking questions
Television	Browsing the Internet
Magazines	Telephone survey
People around us (family, friends, community members, leaders, strangers)	Interview people
	Survey
Shops and businesses	Questionnaire
Experts in specific subjects	

Investigate

3   Look at the items in the photograph.

   a   Explain which of the items in the photograph will be useful if you want to find out:
      i   how many kilometres of main road there are in your country
      ii   whether the inhabitants of your town would like a new resort built on the main beach
      iii   how many cars pass through a particular intersection at busy times
      iv   how many steps you walk in a day.
   b   Explain how each of the other items can be used to collect data.

4   Look at this tally chart and answer the questions.

	Class 5	Class 6
Girls	//// //// //// ////	//// //// //// /
Boys	//// //// //// ///	//// //// //// //

   a   How many boys are in Class 6?
   b   How many girls are in Class 5?
   c   How many more girls than boys are in Class 5?
   d   How many boys are there altogether in the two classes?
   e   How many students are there in Class 6 in total?

5   The table gives you information about different types of electronic media in stock at an electronics store.

	32 GB	512 GB	1 TB
USB flash drives	98	12	0
Memory cards	113	43	5
Portable hard drives	0	35	214

   a   How do you think the store collected this data?
   b   Why is data like this important for a business?
   c   How many portable hard drives are there in stock altogether?
   d   How many more 32 GB memory cards are in stock than 512 GB cards?

6   Data about the food students bought in the school canteen was collected for one day. Draw a tally chart and a frequency table to show the data.

sandwich   cookies   cold drink   cookies   fruit   sandwich   cold drink   cookies
cookies   cold drink   chocolate   fruit   fruit   chocolate   sandwich
sandwich   cookies   cold drink   fruit   cookies   sandwich   cookies

Investigate

Investigate

7 You will now collect data about what
   TV shows people like to watch.

   a Which is your favourite TV show?

   b Choose five TV shows that you think
      are most popular among students in
      your class.

   c Draw up a table like this one and collect
      the data using the method you think
      most suitable.

Favourite TV show	Number of students

   d Write a short paragraph about the data you collected.

8 The school canteen is preparing packed lunches for a class outing.
   The cook wants to find out which sandwiches, cake, fruit and cold drinks
   are most popular so that she can pack these into the lunch parcels.

   Imagine that your class is going on the outing. The canteen offers 4 types
   of sandwiches, 3 types of chips, 5 different fruits and 3 kinds of cold drink.
   Each student can choose any two sandwiches and one of each other item.

   a Write up a questionnaire that will help the cook find the information
      she needs. Decide which kinds of sandwich, chips, fruit and cold drink
      to include as choices. Make sure you provide the correct number of
      choices for each.

   b Use your questionnaire to do a survey in your class to find out what choices students
      would prefer.

   c Draw up a table to organise the data you collected.

   d If the cook puts only the most popular choices in the packed lunches, what will these include?

Investigate

9 Work in groups. Imagine your school wants to encourage students to take part in sports.
   They want to introduce new sports at the school. Decide how you could help the school
   make a decision about which sports to introduce. You will need to collect data and make
   a presentation to back up your suggestions.

What did you learn?

1 Explain what the term 'data' means and where it comes from.

2 Suggest three different ways of collecting data about community issues.

B Averages

Explain

An **average** is a value that is typical of a set of data. There are different types of averages in statistics: the **mean**, the **mode** and the median. This year you will work with the mean and the mode.

To calculate the mean you add all the values of the numbers in a set and then divide the total by the number of values in the set.

Example 1
Find the mean of this data set:

4, 7, 8, 5, 3, 9

$4 + 7 + 8 + 5 + 3 + 9 = 36$ Find the sum of the values

$36 ÷ 6 = 6$ Divide the sum by the number of data items

The mean is 6.

The mode of a set of numbers is the number that appears most often.

Example 2
Work out the mode of this data set:

7, 7, 8, 2, 11, 10

7 occurs most often, so the mode is 7.

A set of data has no mode if there is no value that is repeated. There may also be more than one mode.

Maths ideas

In this unit you will:
* calculate the mean of a data set
* find the mode in a set of data
* use the mean and mode to make sense of a data set.

Key words

average

mean

mode

---

1   Calculate the mean of each set of data.
   a   11, 16, 12, 19, 9
   b   15 mm, 21 mm, 18 mm, 19 mm, 23 mm
   c $2.50, $3.50, $4.00, $7.00, $1.00
   d   36 °C, 37 °C, 30 °C, 32 °C, 34 °C, 33 °C

2   The table shows the number of flies a chameleon caught and ate each day for a week.

Monday	Tuesday	Wednesday	Thursday	Friday	Saturday	Sunday
17	12	14	16	11	14	13

   a   Find the mode of the data. What does the mode tell you?
   b   Calculate the mean number of flies eaten per day.

3   The test marks (out of 20) of ten students are given below.

15	11	18	15	14
19	12	13	14	15

   a   What is the mean mark for this group?
   b   What is the mode?

4   The table shows the amounts five of a group of six children saved.

Name	Anne	Maria	Sharon	Vincent	James	John
Amount saved	$16.50	$18.45	$21.48	$20.13	$15.84	?

   a   Calculate the mean amount the girls saved.

   b   If the mean amount the boys saved is the same as that of the girls, work out how much John saved.

5   A football club needs to choose the best strikers for a competition. They decide to look at the goals scored by each striker in the last 5 games. Here is the set of data.

   a   Which striker had the highest average number of goals? What is the mathematical term for this? What mathematical calculation did you do?

   b   Based on this data, who is the worst striker? Why?

   c   Which strikers should they choose? Why?

Name	Goals
Kirkland	0, 1, 0, 2, 1
Jo	1, 1, 2, 1, 0
Ryan	2, 1, 2, 2, 3
Ahkeem	0, 1, 2, 0, 0
Benoni	1, 0, 0, 0, 1

Investigate

6   Work with a partner. You will need your own calendar page showing at least a month.

   a   Calculate the mean of the numbers in the shaded part of the calendar. What do you notice about the mean?

   b   On your own calendar, mark any other three-by-three set of numbers. Calculate the mean of the numbers you've marked.

   c   Can you predict what the mean will be of any three-by-three set of numbers on a calendar? Check your prediction. Can you explain what happens?

Sun	Mon	Tues	Wed	Thurs	Fri	Sat
				1	2	3
4	5	6	7	8	9	10
11	12	13	14	15	16	17
18	19	20	21	22	23	24
25	26	27	28	29	30	31

What did you learn?

1   Copy and complete the sentences.

   a   The _____ of a set of data is the sum of the values, divided by the number of values in the set.

   b   The mode of a set of data is _____.

2   Use this set of data: $41.00 $79.00 $34.00 $70.00 $34.00 $68.00 $59.00

   Choose the correct answer.

   a   The mean is _____.

$54.00	$45.00	$55.00	$70.00

   b   The mode is _____.

$54.00	There is no mode	$29.00	$34.00

Topic 8 Review

Key ideas and concepts

1   Copy this flow diagram and complete it to summarise how to collect and organise data.

| Create questions | → | | → | | → | |

2   Write short notes to teach someone how to:

    a   find the mode of a set of data   b   calculate the mean of a set of numbers.

Think, talk, write ...

1   Discuss these questions in groups.

What's the difference between a survey and an interview?

Why do you sometimes need to do research and use sources to find data?

How do frequency tables make data easier to work with?

Can data have a mode of 0?

Can the mean of a data set be a number that is not in the set?

2   Write a short paragraph in your maths journal about what you enjoyed most in this topic and why you enjoyed it.

Quick check

1   Sally wants to buy a mobile phone. She finds these prices advertised online in US dollars.

    a   What data did Sally collect?

    b   What is the source of her data?

    c   What is the mean price of the phones advertised?

    d   Can Sally buy one of these phone for the mean price? Explain why or why not.

    e   Which phone price is closest to the mean price?

    f   What is the mode of the prices?

$49.99 $19.99 $25.99 $49.99 $39.99 $45.99 $25.99 $24.99

2   Read these headlines from various news sites on the Internet. Then choose one headline and say how you think the data was collected and organised.

The average adult spends 4 hours 40 minutes per day on their mobile phone.

Most people use their phone to pass the time while they are waiting.

28% of phone users have no lock screen on their phone and only 25% have a PIN code to access the phone.

Teaching notes

Classifying numbers

* ✱ Students have worked with odd and even numbers, square numbers (a number multiplied by itself) and prime numbers (numbers with only two factors, 1 and the number itself).

* ✱ Composite numbers can be written as a product of factors without using 1. So any number that isn't prime is composite – except 1. You can teach students that composite numbers can be represented as rectangles with 2 or more rows. Prime numbers can only be shown using one row.

Factors and multiples

* ✱ Any number that divides into another without remainder is a factor of that number.

* ✱ The highest common factor (HCF) of two or more numbers is the greatest factor shared by the numbers.

* ✱ When you multiply a whole number by any other whole number, the product is a multiple of that number. For example, $5 \times 1 = 5$, $5 \times 2 = 10$, $5 \times 3 = 15$. The products 5, 10 and 15 are all multiples of 5.

* ✱ 10 is the lowest number that is both a multiple of 5 and a multiple of 10, so it is the lowest common multiple (LCM) of the two sets.

Prime factorisation

* ✱ You can write all composite numbers as a product of their prime factors. You can find the prime factors by dividing the composite number by prime numbers. Start with 2 and then move up the prime numbers in order until there is no remainder.

* ✱ Because each number has a unique product of prime factors, you can use the products to find the HCF and LCM of numbers without making long lists. The HCF is the product of the common prime factors, and the LCM is the product of the biggest group of each prime factor from both sets.

Roman numerals

* ✱ Last year students learnt that the Roman number system used letters to represent numbers and that it did not use place value.

* ✱ Use clock faces with Roman numerals to remind them how to write these.

Sharon is playing a game on her computer. The numbers in the circles move quickly across the screen and she has to drag them into one of the boxes before they disappear. Where should she put each number? Think of one more number that could go into each box.

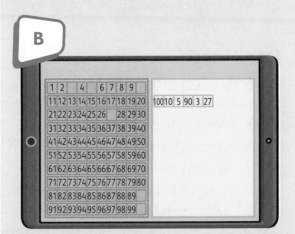

In this game you have to move the blue numbers to the right-hand box to make the longest possible path of numbers that are either a factor or a multiple of the number before it. Play on paper – see who can make the longest chain. Each number can be used once only.

1234

C

Roman Date Converter

XVI·IV·
MMXIX

15	March	2018
16	**April**	**2019**
17	May	2020

Look at the date on this phone. What do you call this type of numeral? How do you write numbers using letters in this system? Can you work out what today's date would be on the phone?

Think, talk and write

A   Classifying numbers (*page 100*)

Here is a set of numbers:

| 2 | 5 | 7 | 9 | 12 | 15 | 16 |
| 20 | 25 | 36 | 80 | 81 | 97 | 100 |

1   List the odd numbers.

2   List the even numbers.

3   List the prime numbers.

4   List the square numbers.

5   What does it mean if we say a number is composite?

B   Factors and multiples (*pages 101–103*)

1   Work in pairs to write a definition and give an example of each of these mathematical terms.
   a   Factor
   b   Multiple
   c   Prime number
   d   Prime factor
   e   Product of prime factors

2   What number is a factor of every whole number above zero?

3   How would you describe each of these sets of numbers? Why?
   a   1, 2, 3, 4, 6, 8, 12, 24
   b   3, 6, 9, 12, 15, 18

4   If Jaime's age is a multiple of 2 and also a square number less than 25, how old could he be?

C   Roman numerals (*page 104*)

Sherie is IX years old and Matt is XI years old.

1   What are their ages?

2   Explain how the Romans used the X and the I to make two different numbers.

3   What is the sum of the ages?

4   How would you write the answer in Roman numerals? Explain your answer.

A Classifying numbers

Explain

We can classify numbers using their properties.

Even numbers are all divisible by 2. They have 0, 2, 4, 6 or 8 in the ones place.

Odd numbers are not divisible by 2. They have 1, 3, 5, 7 or 9 in the ones place.

A number multiplied by itself (for example, 6 × 6) produces a **square number** (36) and the original numbers (6) are called the **square root**.

A **prime number** is one that is only divisible by 1 and itself (it has only two factors).

A **composite number** is a number that has more than two factors.

Maths ideas

In this unit you will:
* use the terms odd, even, prime, square and composite to classify numbers in different ways.

Key words

even numbers
odd numbers
square number
square root
prime number
composite number

1 Write each set of numbers.
   a Odd numbers between 120 and 150
   b Even numbers greater than 40, but less than 60
   c The first six prime numbers
   d Composite numbers between 20 and 30

2 Say whether these statements are true or false.
   a All prime numbers are odd.
   b 21 is a prime number.
   c 2 is both square and composite.

3 Use the digits 1, 4, 7 and 8 once only to make:
   a the greatest possible even number
   b the smallest possible odd number
   c a four-digit prime number.

Investigate

4 Work with a partner. You can use a calculator in this investigation.
   The largest known prime number was discovered in January 2018. It has 23.2 million digits.
   a Write out 100 digits and time how long it takes. Use that to try and estimate how long it would take to write this number.
   b A computer would print about 3 500 digits per sheet of paper. About how many sheets of paper would you need to print out this number?

What did you learn?

Classify each of these numbers as fully as possible.

| 23 | 45 | 64 | 91 | 100 |

B Factors and multiples

Explain

Read through the information and the examples to revise what you learnt about factors and multiples last year.

The **factors** of a number are all the whole numbers that divide into it exactly.

The factors of 12 are 1, 2, 3, 4, 6 and 12.

Some numbers share factors. Shared factors are called common factors.

The purple factors are common to 12 and 16.

Factors of 12: **1**, **2**, 3, **4**, 6, 12

Factors of 16: **1**, **2**, **4**, 8, 16

The **highest common factor (HCF)** of 12 and 16 is 4.

Multiples of a number are found when you multiply that number by any other whole number.

The first ten multiples of 3 are: 3, 6, 9, 12, 15, 18, 21, 24, 27, 30.

The first ten multiples of 4 are: 4, 8, 12, 16, 20, 24, 28, 32, 36, 40.

12 and 24 are multiples of both 3 and of 4. We call these common multiples.

The lowest number that is common to both sets is 12.

We say that 12 is the **lowest common multiple (LCM)** of 3 and 4.

Prime numbers are numbers with two (and only two) factors. The factors are 1 and the number itself.

7 is a prime number. The only two whole numbers that can divide exactly into 7 are 1 and 7 itself.

The factors of 7 are 1 and 7.

A factor that is a prime number is called a prime factor.

All numbers can be written as a **product of prime factors**.

To find the prime factors of a number you can use a prime factor tree.

Maths ideas

In this unit you will:
* find the lowest common multiple (LCM) and highest common factor (HCF) of sets of numbers
* write numbers as a product of their prime factors.

Key words

factors
highest common factor (HCF)
multiples
lowest common multiple (LCM)
product of prime factors

Example

Write 60 as a product of prime factors.

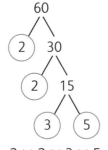

Write the number as a product of any two factors

Circle any prime factors

Write any non-prime factors as a product of two numbers

Continue until all factors are prime

$2 \times 2 \times 3 \times 5$

$= 2^2 \times 3 \times 5$ Write the product using multiplication signs or powers

1. List the first ten multiples of these numbers.
   a 4   b 8   c 7   d 9   e 11

2. What is the LCM of each set of numbers?
   a 4 and 6   b 3, 4 and 8   c 12 and 8   d 2, 5 and 10

3. Find all the factors of these numbers.
   a 32   b 16   c 80   d 100   e 36

4. By listing the factors, find the HCF of each set of numbers.
   a 12 and 16   b 18 and 40   c 40 and 60   d 38 and 39

5. Express the following numbers as a product of prime factors.
   a 14   b 32   c 40   d 36   e 100
   f 156   g 225   h 80   i 24   j 1 000

Problem-solving

6. Sally, James and Keshorn are checking their email accounts to see if an invitation to a party has arrived. Sally checks hers every 4 minutes, James checks his every 6 minutes and Keshorn checks his every 14 minutes.
   a After how many minutes will Sally and James check their accounts at the same time?
   b After how many minutes will all three of them check their accounts at the same time?

7. Toniqua has 60 red beads and 90 white beads. She wants to use all the beads to make identical bangles. Each bangle should have the same number of red and the same number of white beads. What is the greatest number of bangles she can make?

8. Nicky has two lengths of ribbon. One is 32 cm long and the other is 60 cm long. She wants to cut as many pieces of equal length as possible from the two ribbons. How long will each piece be?

Explain

It can take a long time to list multiples or work out factor pairs to find the lowest common multiple or highest common factor.

Writing a number as a product of its prime factors saves time and allows you to find the LCM and HCF efficiently.

Example 1

Find the HCF of 24 and 60.

Use factor trees or division to find the prime factors.

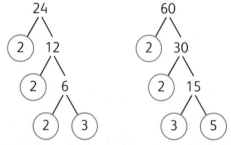

$24 = ②×②× 2 ×③$
$60 = ②×②×③× 5$

Mark the common factors
$2 × 2 × 3$ is common to both sets

Find the product of the common prime factors: $2 × 2 × 3 = 12$
The HCF is 12.

Example 2

Find the LCM of 8 and 14.

Use factor trees or division to find the prime factors.

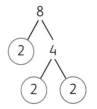

$8 = ②\times②\times②$ Find where each different factor appears most often and mark that factor each time it appears in that set.

$14 = 2 \times ⑦$

Find the product of the marked factors: $2 \times 2 \times 2 \times 7 = 56$

The LCM is 56.

9   Use prime factors to find the HCF of each pair of numbers.

a 15 and 20	b 12 and 16	c 20 and 16	d 30 and 36
e 25 and 35	f 10 and 20	g 18 and 30	h 36 and 63

10   Write each number as a product of its prime factors and write the HCF of each set.

a   75, 120 and 150   b   24, 40 and 80   c   12, 48 and 60

11   Use prime factorisation to find the LCM of each pair of numbers.

a 15 and 18	b 24 and 28	c 22 and 25	d 18 and 24
e 25 and 30	f 72 and 108	g 95 and 120	h 22 and 33

12   Determine both the HCF and the LCM of each pair of numbers using prime factorisation.

a   36 and 60   b   36 and 48   c   60 and 80   d   52 and 78

What did you learn?

1   a   Marcus has written some incorrect factors in each list. Which are they?

> Factors of 24: 4, 2, 6, 8, 1, 16, 24, 3
> Factors of 23: 3, 1, 7, 23

   b   How can working in order with factor pairs help you to avoid mistakes like Marcus made?

2   a   Find the LCM of 2, 3, 4 and 6.

   b   What is the HCF of 2, 3, 4 and 6? Why?

3   Use prime factors to find the LCM and HCF of 8, 12, 16 and 30.

C Roman numerals

Explain

You already know that **Roman numerals** are written using letters to represent numbers.

I = 1	V = 5	X = 10	L = 50
C = 100	D = 500	M = 1 000	

The Roman number system does not use place value.
The letters are combined in different ways to make numbers.

Maths ideas

In this unit you will:
* read and write Roman numerals.

Key words

Roman numerals

Example 1
Add the values when the letters are the same:
III = 1 + 1 + 1 = 3
XXX = 10 + 10 + 10 = 30
MM = 1 000 + 1 000 = 2 000

Example 2
Add the values when letters of smaller value are to the right of a letter of greater value:
VII = 5 + 1 + 1 = 7
XXV = 10 + 10 + 5 = 25
CCLX = 100 + 100 + 50 + 10 = 260

Example 3
Subtract the values when letters of smaller value are to the left of a letter of greater value:
IX = 10 − 1 = 9
XL = 50 − 10 = 40
CM = 1 000 − 100 = 900

Example 4
When a letter of smaller value is between two letters of greater value, subtract it from the greater value to the right.
XIV = 10 + (5 − 1) = 10 + 4 = 14
LVL = 50 + (50 − 5) = 50 + 45 = 95

1 Write these numbers using the Roman number system.
   a 12   b 10   c 9   d 19   e 20   f 17

Problem-solving

2 What is the value of each of these Roman numbers?
   a XXX   b DCC   c CCM
   d LM   e LVL   f MIX
   g DCLXIII   h XIV   i MMIX

3 Write the present year using Roman numerals.

4 Use only the letters C and D to make the smallest and the greatest Roman number possible.

What did you learn?

Count in 5s from 0 to 100. Try to write the numbers you count using Roman numerals.

Topic 9 Review

Key ideas and concepts

Write down the correct mathematical term for each statement and give an example to show what each one means.

1 Numbers that have only two factors

2 Numbers that cannot be divided by 2 with no remainder

3 A number that will divide into another with no remainder

4 The highest number common to sets of factors

5 Numbers with 0, 2, 4, 6 or 8 in the units position

6 The lowest number in two or more sets of multiples

7 A number with more than two factors

8 The product of a number and itself

9 A method of writing a number using prime factors

Think, talk, write ...

These statements are all *incorrect*. Discuss why they are incorrect.

1 The LCM of two numbers is the lowest number that divides into them both with no remainder.

2 To find the HCF of two numbers you just multiply them.

3 The LCM of 4 and 6 is 24, because $4 \times 6 = 24$.

4 The highest common factor of two prime numbers is the sum of the numbers.

Quick check

1 Say whether each statement is true or false.

   a Common multiples of 3 and 5 must be odd numbers.

   b All common multiples of 2 and 5 are also multiples of 10.

   c The HCF of 5 and 11 is 1.

   d 34 has four ordinary factors and two prime factors.

   e The smallest number you can find with exactly three factors is 9.

   f XX is the Roman number for 22.

   g All square numbers are also composite numbers.

2 List:

   a all the factors of 36

   b the prime factors of 36

   c the four lowest multiples of 36

   d the factors of 36 that are square numbers.

3 Find the difference between the sum of the prime numbers less than 10 and the sum of the prime numbers between 10 and 20.

4 Use prime factors to find the HCF and LCM of 12, 15 and 18.

Topic
10 Algebraic thinking

Teaching notes

Special sequences

* A sequence is a set of numbers that obeys a particular rule. For example, 2, 4, 6, 8, … is the sequence of even numbers, and 1, 4, 9, 16, 25, … is the sequence of square numbers.

* Some sets of numbers are mathematically recognised as special sequences. Square numbers are one of these sequences. So are triangular, rectangular and oblong numbers. There are algebraic rules for working with these sequences, but at this level, students use arrays and dot diagrams to represent the numbers and classify them.

Using equations to solve problems

* Writing a number sentence with one or more unknown quantities is a useful problem-solving strategy. In Level 5, students worked with letters (such as x) to represent unknowns. The unknown letters are called variables in mathematics.

* Number sentences with an equals sign (=) are called equations. The quantities on either side of the sign must be equal. This means that if you perform an operation on one side of the equals sign, you must do the same operation on the other side to maintain the equality. Revise solving basic equations with the class if you feel they need practice before you move on to the new concepts. You can find a number of online sites that teach the ideas and offer interactive activities.

* Make sure students understand that they can use any letter when they set up their own equations, but they must state what the letters represent.

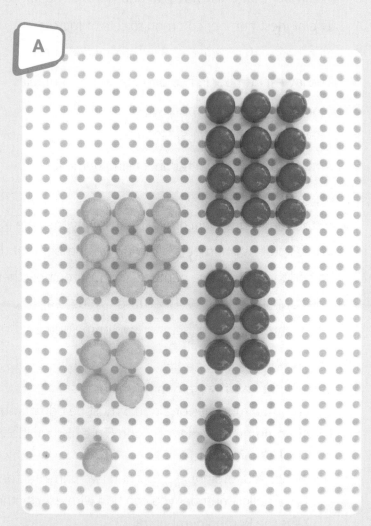

Look at the two dot patterns on this pegboard. Do you recognise either of them?
How many pegs would you need to build the next shape in each pattern?
How did you work this out? How can you tell from this that the first two numbers in each pattern are not composite numbers?

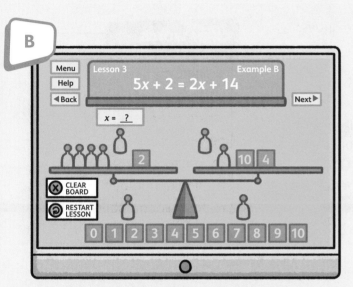

A teacher uses a computer game to teach her class about equations. One blue weight on the balance scale represents x. You can move them and swap numbers around. Work in pairs. What steps would you follow to show the students how to find the value of x in this example? Why?

Think, talk and write

A **Sequences** *(pages 108–110)*

1 The pattern of odd and even numbers is one of the most basic sequences in mathematics.
   a List the odd numbers between 40 and 60.
   b List the even numbers between 100 and 116.
   c Explain how you can tell whether a number in the millions is odd or even.

2 A square number is the product of multiplying a number by itself, for example, $3 \times 3 = 9$. So, 3 squared equals 9 and we say that 9 is a square number. When you draw dot diagrams of square numbers you get a sequence of squares.
   a These dot diagrams show some other numbers. What would you call these numbers? Why?

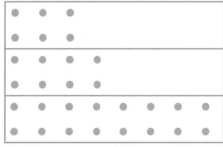

   b How could you work out each number in the sequence without counting all the dots?

B **Using equations to solve problems** *(pages 111–112)*

Write an equation to represent each problem and then solve it to work out the unknown value.

1 15 more than x is 20
2 10 less than a is 30
3 15 minus b is 10
4 Half of x is 22
5 Twice y is 30
6 The square of x is 16
7 The sum of an unknown number and 12 is 23

A Sequences

A **sequence** is a set of numbers that obeys a particular rule or has a certain **pattern**. For example, 2, 4, 6, 8, … is the sequence of even numbers.

Some numbers can be represented geometrically using shapes made from dots.

Square numbers can be represented using square arrays.

When you represent a number as a square array like this, the number of dots on one side of the square is multiplied by itself to find the total number of dots.

The total number of dots is the square number or **perfect square**. The length of the side is the **square root**.

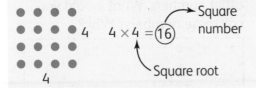

$4 \times 4 = 16$ → Square number

Square root

Maths ideas

In this unit you will:
* investigate number patterns and learn about special sequences of numbers.

Key words

sequence
pattern
square numbers
perfect square
square root
triangular numbers
rectangular numbers
oblong numbers

Triangular numbers can be represented using dots to form triangles. The triangles can be isosceles or right-angled.

Both of these sets of dots show the first four triangular numbers.

There is a difference of 1 between the number of dots in each row or column.

Each triangular number can be written as a sum of consecutive numbers.

1 + 2 = 3
1 + 2 + 3 = 6
1 + 2 + 3 + 4 = 10, and so on.

Rectangular numbers can be shown as rectangles with more than one row and column.

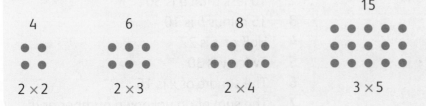

4	6	8	15
2×2	2×3	2×4	3×5

A rectangular number is the product of its length and width.

Rectangular numbers can be written as a product of two factors.

All rectangular numbers are composite numbers. They cannot be prime numbers but they can be square numbers.

Oblong numbers are a special type of rectangular number. These numbers can be shown as rectangles with a difference of 1 between the length and breadth.

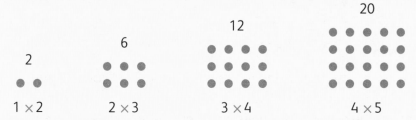

Oblong numbers can be written as the product of two consecutive whole numbers.

$0 \times 1 = 0$

$1 \times 2 = 2$

$2 \times 3 = 6$

$3 \times 4 = 12$

$4 \times 5 = 20$

There is one prime oblong number. Can you see it above?

1   Without drawing them, write down the 8th, 10th and 20th square numbers.

2   Look at this set of numbers.

3	5	7	8	12	30	17	11
36	15	100	40	19	25	21	12

   a   Which of these are square numbers?

   b   Find two triangular numbers in the set.

   c   Is 7 a rectangular number? Why?

   d   Draw a diagram to show that 30 is both rectangular and oblong.

   e   Which of the numbers are not square, triangular, rectangular or oblong?

Problem-solving

3   Micah built a number using counters. It is an oblong number. One side of the shape is 11 counters long. What could the other side be? Explain your answer.

4   Jessica says that 21 cannot be a rectangular number because it is a triangular number. Is she correct? Explain your answer.

5   Sharon is playing around with triangular numbers. She thinks that all triangular numbers can be rearranged to make rectangles. She tries with 15 and it works.

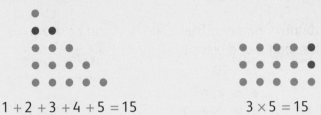

$1 + 2 + 3 + 4 + 5 = 15$ $3 \times 5 = 15$

a   Does this method work for the next two triangular numbers – 21 and 27? Draw diagrams to check your answer.

b   Sharon says that you can group the number into pairs and then multiply to find the answer faster when you are working with triangular numbers. She writes these notes in her book:

> $1 + 2 + 3 + 4 + 5 + 6 + 7 + 8 + 9 + 10$
>
> $= (1 + 10) + (2 + 9) + (3 + 8) + (4 + 7) + (5 + 6)$
>
> This is the same as $11 \times 5 = 55$
>
> If you work out the sum of each pair and how many there are,
>
> you do not need to write them all out.

Test Sharon's idea. Does it work?

c   Royston looks at the first five triangular numbers and he sees a pattern.

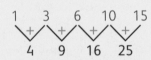

Look at his diagram and explain mathematically what he has seen.

Do you think this will work for all triangular numbers? Give reasons for your answer.

What did you learn?

1   Show that 6 is a triangular, rectangular and oblong number.

2   What is the fewest number of counters that you can move to show that this triangular number is also square? You can use real counters to help if you need to.

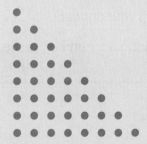

B Using equations to solve problems

Explain

You already know that an **equation** is a mathematical statement that makes two quantities equal. In other words, it is a **number sentence** with an equals sign.

$2 + 7 = 9$ is an equation. There are no unknown values in this equation.

$2 + x = 12$ is also an equation. The letter x is a **variable** that represents an **unknown quantity**.

You can **solve** the equation by working out the value of x. (You should be able to see that $x = 10$ in this equation.)

Equations are very useful as a problem-solving strategy.

When you have a word problem, you can choose variables (letters) to represent unknown amounts and write an equation to represent the problem.

Maths ideas

In this unit you will:
* write number sentences using letters to represent unknown values
* use equations to solve problems.

Key words

equation
number sentence
variable
unknown
quantity
solve

It is important to say what the variable represents when you set out your work.

Read through these examples to see how to set up and solve equations to solve problems.

Example 1

The product of a number and 7 is 42. What is the number?

Let the number be x.	Say what the variable represents
$7 \times x = 42$	Write an equation
$x = 6$	Solve the equation (divide both sides by 7)
$\therefore$ The number is 6.	Write an answer statement

Note

$\therefore$ is a mathematical symbol that means therefore.

Example 2

If a number is multiplied by 2, and then 5 is added to the product, the result is 21. What is the number?

Let the number be y.	Say what the variable represents
$(y \times 2) + 5 = 21$	Write an equation – this one has two steps
$(y \times 2) + 5 - 5 = 21 - 5$	Subtract 5 from both sides
$y \times 2 = 16$	Solve: divide both sides by 2
$y = 8$	Check $8 \times 2 + 5 = 16 + 5 = 21$
$\therefore$ The number is 8.	Write an answer statement

Example 3

After buying a packet of chips, Andy has 25¢ left from $1.00. How much did the chips cost?

Let the cost of the chips be p.	
$1.00 = 100¢$	Change dollars to cents so you can work with the same units
$p + 25 = 100$	Write an equation (this one could also be $100 - p = 25$)
$p + 25 - 25 = 100 - 25$	Solve the equation: subtract 25 from both sides
$p = 75$	Check: $75¢ + 25¢ = 100¢ = \$1.00$
$\therefore$ The chips cost 75¢.	Write an answer statement

1   For each problem, write an equation and solve it, even if you can see the answer without doing any working.

    a   The sum of a number and 6 is 23. What is the number?

    b   A number less 9 is 42. What is the number?

    c   A number less 7 is 51. What is the number?

    d   The product of a number and 8 is 72. What is the number?

    e   When a number is divided by 5, the quotient is 8. What is the number?

    f   Half of a number is added to 8 to get 23. What is the number?

2   Solve these equations. Show at least two steps in each one.

    a $2 \times x + 10 = 25$ b $x \times 10 - 12 = 18$ c $(x + 10) \times 5 = 75$

    d $3 \times x - 15 = 12$ e $5 \times x - 100 = 400$ f $x \times 5 + 14 = 44$

> **Problem-solving**

3   Lamar bought a pen and a sharpener for $2.25. If the pen cost $1.50, what did the sharpener cost?

4   It takes Yohan 45 minutes to do his mathematics and science homework. If the science took him 23 minutes, how long did the mathematics take?

5   Kezia bought tickets for a show for herself and her friends. The tickets cost $5.00 each and she paid a total of $105.00. How many tickets did she buy?

6   Nadia, Ann and Kim have a combined age of 27. If Nadia and Ann are the same age, and Kim is 14, how old are Nadia and Ann?

7   Jerome put 156 counters into equal groups of 12. How many groups could he make?

8   The perimeter of each shape is given. Work out the length of each unknown side.

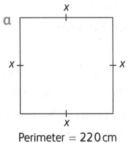

a
Perimeter = 220 cm

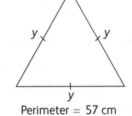

b
Perimeter = 57 cm

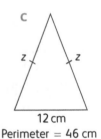

c
12 cm
Perimeter = 46 cm

9   Tonya got 15¢ change from $1.00 when she bought 7 plums. How much did each plum cost?

10   A mango costs 10¢ more than an orange. Together, a mango and an orange cost 98¢. What does each fruit cost?

11   Bruce weighs 6 kg less than Keshawn. If the sum of their masses is 88 kg, how much does each boy weigh?

What did you learn?

1   What number must be added to 7 so that when the result is doubled, the answer is 22?

2   The length of a rectangle is 4 cm more than its width. If the perimeter is 16 cm, how long are the sides?

Topic 10 Review

Key ideas and concepts

For each term below, write a definition in your own words and give an example to show what it means.

pattern	sequence	rectangular number
square number	triangular number	equation

Think, talk, write …

1   Look at this ornament. What types of numbers did the designer need to use to make this?

2   How can you check whether the solution to an equation is correct or not? Discuss your ideas with your group.

Quick check

1   Choose numbers from the box to answer the questions.

12   20   14   15   21   3   1   2   11   25   100

Which numbers are:

a   oblong?

b   square?

c   prime?

d   triangular?

e   oblong and rectangular?

2   Say whether these statements are true or false. Give a reason for your answer.

a   There are no prime square numbers.

b   All oblong numbers are also rectangular numbers.

c   A rectangular number can also be a square number.

d   All triangular numbers are also rectangular.

e   A square number can also be oblong.

3   Write each of these as a mathematical equation and find the value of n.

a   The sum of n and 5 is 23.

b   The sum of n and 9 is 17.

c n less 12 results in 34.

d   The product of n and 12 is 96.

e   The sum of half n and 13 is 63.

4   A number less 9 is equal to the product of 5 and 8. What is the number?

5   If a number is multiplied by 10, and then 6 is added to the result, the answer is 96. What is the number?

Teaching notes

Perimeter

* Perimeter is a linear measure of the total length of the boundaries (sides in polygons) of a 2-D shape.
* You can measure perimeter, but you can also calculate perimeter if you are given the lengths of the sides.
* To calculate perimeter, you find the sum of the lengths of all sides of the shape.
* You can use formulae to find the perimeter of some shapes, but in the case of irregular polygons it is often quicker to add the lengths of the sides.

Area

* The area of a 2-D shape is the amount of space it covers.
* Standard units of area are called square units. A square with sides of 1 cm makes 1 square centimetre (1 cm²).
* Larger areas can be measured in square metres or square kilometres.
* The area of many shapes can be calculated using a formula. For example, the area of a rectangle with a length of 4 cm and a width of 2 cm can be calculated using this formula: area = length × width, so 4 cm × 2 cm = 8 cm².
* This year students will use the fact that a right-angled triangle is half of a rectangle with the same length and breadth to derive the formula for the area of triangles: $A = \frac{1}{2}$ length × breadth. This formula applies to all triangles, but the focus at this level is on right-angled triangles only.
* To find the area of composite shapes, you can divide them into smaller areas and then add these, or you can calculate the area of a larger shape and subtract areas from it. The method you use depends on the shape you are dealing with.

Volume

* This is a new concept for students. Volume is the amount of space that a solid object (or an amount of liquid) occupies. Experiment with cubes to demonstrate this to the class, if possible.
* Volume is measured in cubic units (such as cm³). Remind the students that when you multiply cm by cm you get cm². Similarly, when you multiply cm × cm × cm, you get cm³.

A

Each of these solar panels has an aluminium frame around it. How can you work out the total length of the aluminium without measuring each one? What information would you need about each panel to do this?

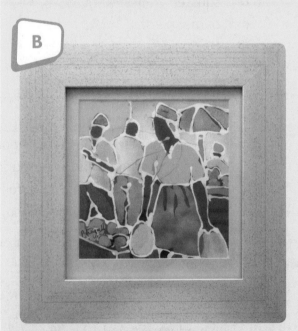

B

Look at this framed artwork from Barbados. Estimate which is greater – the area covered by the artwork or the area covered by the wooden frame? Tell your partner how you decided.

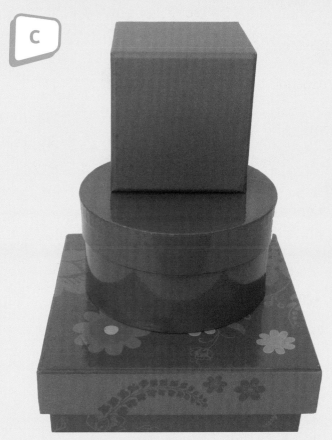

Which box do you think takes up the most space? Which box do you think could hold the most sand? How did you decide? Discuss your methods with your group.

Think, talk and write

A   Perimeter *(pages 116–117)*

1   Sean ran round the outside of this field three times.

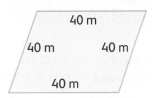

    a   How far did he run each time?

    b   How far did he run altogether?

2   Estimate the perimeter of the front of the door in your classroom. Then measure it. How accurate was your estimate?

B   Area *(pages 118–120)*

1   Which of these could be measurements of area? How do you know?
$16\ cm^2$ $100\ m^2$ $18\ cm$ $12\ m^2$

2   What do you need to measure if you want to cover a bathroom wall with tiles? How can you work out how many tiles you need?

3   Each square on this diagram represents 1 cm squared. What is the area of each shape?

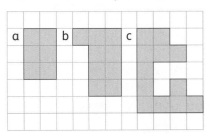

C   Volume *(pages 121–122)*

Look at these solids. How many blocks would you need to build each one?

A Perimeter

The **perimeter** of a shape is the **distance** around it.

To **measure** the actual perimeter, you need to measure the **length** of each **side** and add up these lengths. If you know the lengths of the sides, you can simply **calculate** the **sum**.

To measure a curved side, you can lay a piece of string along the edge and then measure the length of the string.

You will use mm, cm, m or km to measure perimeter, depending on the size of the shape that you are measuring.

You can use formulae to calculate the perimeter of some plane shapes.

Rectangles
The opposite sides (length and width) are equal, so the perimeter (P) equals twice the length (*l*) plus twice the width (*w*)
$P = 2 \times l + 2 \times w$ or $P = 2 \times (l + w)$

Squares
All four sides are the same length, so
the perimeter (P) equals four times the length of one side (*s*)
$P = 4 \times s$

Maths ideas

In this unit you will:
* measure and calculate the perimeter of shapes
* solve problems involving perimeter.

Key words

perimeter
distance
measure
length
side
calculate
sum

1   Measure the boundaries of each shape in millimetres or centimetres and calculate the perimeter.

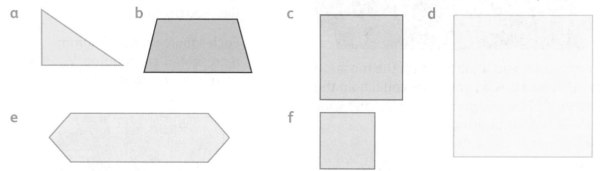

a   b   c   d

e   f

2   Estimate and then measure the perimeter of these shapes in your home or at school.
   a   A window   b   A carpet or rug
   c   The screen on an electronic device   d   A sports field

3   Use the measurements that are given to calculate the perimeter of each shape.

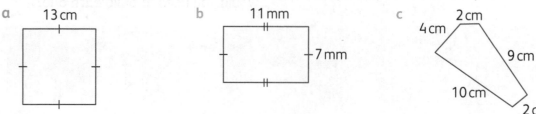

   a   13 cm   b   11 mm   c   2 cm
                                                                  4 cm
                                              7 mm   9 cm

                                                                  10 cm
                                                                          2 cm

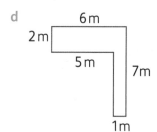

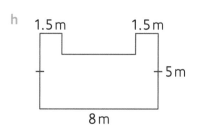

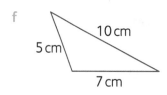

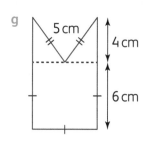

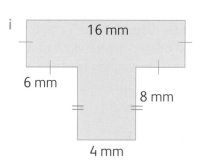

4   Draw the following shapes.
   a   Two different triangles, each with a perimeter of 12 cm
   b   Two different rectangles, each with a perimeter of 24 cm

Problem-solving

5   a   The perimeter of a square table cloth is 5 m. What size square will this cloth cover?
   b   The perimeter of a rectangle is 20 cm. One side is 8 cm. How long is the other side?
   c   Tori-Anne wants to put patterned tape around the perimeter of the front of her exercise book. The book measures 15 cm × 21 cm. The tape is 1 cm wide. How many centimetres of tape will she need to cover the perimeter?
   d   Seymour wants to put some photographs on a poster. The poster is 15 cm × 30 cm. He wants to put at least six photographs on the page. He wants a little space around each photograph. What should the perimeter of the photographs be?
   e   One side of a square lawn measures 10 m. What is the perimeter of the lawn?
   f   The width of a rectangular field is 8 m. Its length is twice the width. What is the perimeter of the field?

What did you learn?

1   Which of these is the formula for calculating the perimeter of a rectangle?

$$P = 2 \times (l + w) \qquad P = 4 \times l + 4 \times w$$

2   Write the formula for calculating the perimeter of a square.

3   Calculate the perimeter of each shape.
   a   An equilateral triangle with one side measuring 7.5 cm
   b   A square with sides measuring 3.1 cm
   c   A rectangle that measures 2.5 m × 3 m
   d   A pentagon with side lengths of 12 mm, 11 mm, 15 mm, 15 mm and 14 mm

117

B Area

Area is a measure of how much space a flat shape takes up. You measure area in **square units** such as **square centimetres (cm²)**. One square centimetre (1 cm²) is a square with a **length** and **width (breadth)** of 1 cm. One **square metre (m²)** is a square with a length and width of 1 m. You calculate area by counting square units or by multiplying.

Last year you worked out **formulae** to calculate the area of rectangles and squares.

Rectangles

The area (A) equals the length (*l*) times the width (*w*):

$A = l \times w$

Squares

The sides of a square are equal.

So the area (A) equals the length (*l*) times the length (*l*):

$A = l \times l$ or $A = l^2$

Triangles

There is a special formula for working out the area of a triangle. Look at the diagram.

The area of rectangle ABCD = 4 cm $\times$ 3 cm = 12 cm².

The right-angled triangle BCD takes up half the area of the rectangle.

The area of the triangle is half the area of the rectangle, so: $A = \frac{1}{2} \times (l \times w)$

This is the same as $A = (l \times w) \div 2$

In this unit you will:
* revise what you learnt about area and square units of measurement in earlier levels
* use formulae to work out the area of rectangles, squares and right-angled triangles
* solve problems involving irregular areas.

area
square units
square centimetres (cm²)
length
width (breadth)
square metre (m²)
formulae

1 Use a formula to work out the areas of these rectangles and squares.

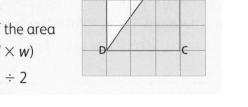

a 5 cm 3 cm
b 7 cm 2 cm
c 4 cm 4 cm
d 5.5 cm 5.5 cm

2 Calculate the area of each shape.

   a A wall that measures 2.2 m by 3.6 m

   b A book cover with a perimeter of 70 cm and a width of 15 cm

   c A billboard with a perimeter of 24 m and a height of 4 m

   d A mobile phone with a width of 6.5 cm and a height that is twice the length of the width

3   Estimate the area of the triangles below by counting the squares. Then use a formula to check your estimates.

a
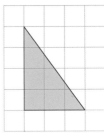

b
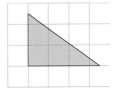

c
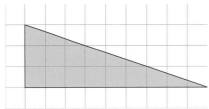

4   Work out the area of each shape.

a
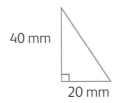

40 mm

20 mm

b
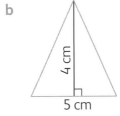

4 cm

5 cm

c
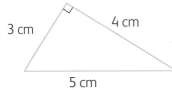

3 cm   4 cm

5 cm

d
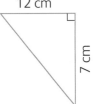

12 cm

7 cm

e
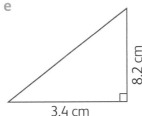

8.2 cm

3.4 cm

f
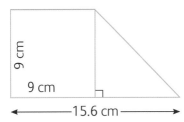

9 cm

9 cm

15.6 cm

Explain

Composite shapes are made by combining shapes or by removing part of a shape.

You can find the area of composite shapes in different ways.

*   You can divide the shape up into known shapes and calculate the area of each one. Then you add all the areas to find the total area.

*   You can work out the area of one shape and subtract the area of any parts cut from it.

Example 1

What is the area of this shape?

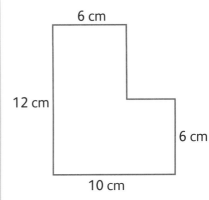

6 cm

12 cm

6 cm

10 cm

Start by dividing the given shape into two rectangles. There are different ways of doing this. Here are two ways.

Method 1   **Method 2**

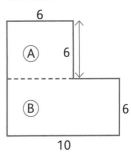

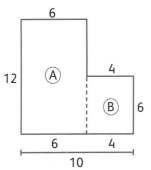

Now work out the area of each rectangle:

Method 1	**Method 2**
$A = l \times w$	$A = l \times w$
Rectangle A = 6×6 = 36 cm²	Rectangle A = 12×6 = 72 cm²
Rectangle B = 10×6 = 60 cm²	Rectangle B = 4×6 = 24 cm²
Area A + Area B = 36 + 60 = 96 cm²	Area A + Area B = 72 + 24 = 96 cm²

Example 2

What is the area of this shape?

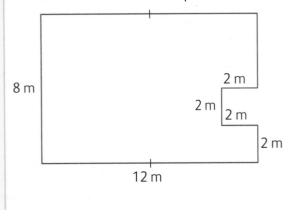

You can divide this shape into three rectangles to find the area, but you can do fewer calculations if you think of this shape as a rectangle with a square removed.

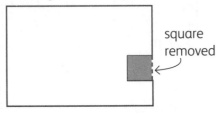

square removed

Area of rectangle = 8×12 = 96 m²
Area of square = 2×2 = 4 m²
Area of shape = 96 − 4 = 92 m²

5   Calculate the area of each shape. Draw a rough diagram in your book to show how you divided up each shape.

a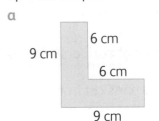

b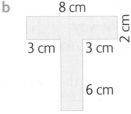

c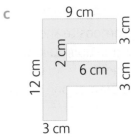

d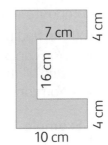

Challenge

6   What are the side lengths of a rectangle with an area of 100 cm² and a perimeter of 50 cm?

What did you learn?

1   Write down these formulae.

    a   A formula to calculate the area of a square

    b   A formula to calculate the area of a rectangle

    c   A formula to calculate the area of a right-angled triangle

2   Work in pairs. Make up problems involving the area of composite shapes for your partner to solve.

C Volume

Volume is the amount of space that an object occupies.

Look at this object. It is a solid made of 4 layers of 4 cubes each.

It takes up 16 cubes of space. We say it has a volume of 16 cubes.

We measure volume in **cubic units**. A **cubic centimetre** is written cm^3, and a **cubic metre** is written m^3.

We calculate the volume of a cube or cuboid using the following formula:
Volume = length $\times$ breadth $\times$ height
$V = l \times b \times h$

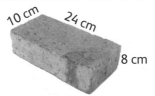

Example

This brick is 10 cm wide, 24 cm long and 8 cm high. Calculate the volume of the brick.

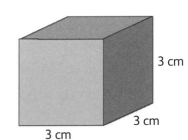

Volume = length $\times$ breadth $\times$ height
        = 24 cm $\times$ 10 cm $\times$ 8 cm
        = 1 920 cm^3

Maths ideas

In this unit you will:
* learn how to work out the volume of solids such as cubes and cuboids
* solve problems involving volume.

Key words

volume
cubic units
cubic centimetre
cubic metre

Think and talk

Solids are three-dimensional objects.

What are the three dimensions of a solid such as a cuboid?

1   Work out the volume of each stack of cubes.

   a   b   c   d   e

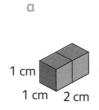
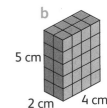
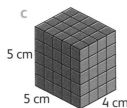
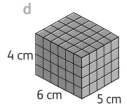
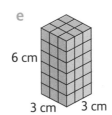

   a: 1 cm, 1 cm, 2 cm
   b: 5 cm, 2 cm, 4 cm
   c: 5 cm, 5 cm, 4 cm
   d: 4 cm, 6 cm, 5 cm
   e: 6 cm, 3 cm, 3 cm

2   Calculate the volume of each of these solids.

   a   b

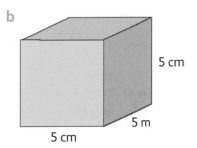

   a: 3 cm, 3 cm, 3 cm
   b: 5 cm, 5 cm, 5 cm, 5 m

Remember that when you have to solve problems involving shape and space, it is useful to draw a diagram and label it to show what you know and what you need to work out.

For problems that involve manipulation of shapes, you can use real objects and build models to help you solve the problem.

Problem-solving

3   A rectangular box is 8 cm long, 6 cm wide and 3 cm high. What is the volume of the box?

4   The volume of a box is 162 cm³. If the length is 9 cm and the width is 6 cm, what is the height of the box?

5   The volume of a cube is 2 cm³. How many of these cubes can fit into a box that is 8 cm long, 5 cm wide and 4 cm high?

6   The volume of a tank is 2 160 cm³. Its length is 15 cm. If the width and the height are equal measurements, what is the width?

Investigate

7   Nadia has a number of small jewellery boxes with a volume of 8 cm³. She wants to pack these into a larger box like the one shown.

   a   Investigate how many small boxes she can fit into the larger box.

   b   Tell your group how you worked out your answer.

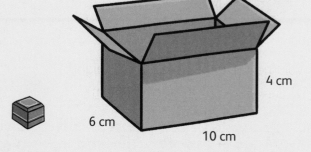

4 cm

6 cm

10 cm

8   Mr Johnson is going to make wooden boxes. Each one must have a volume of 2 000 cm³ and one of the dimensions has to be 25 cm.

   a   What could the other dimensions be? Find as many possible answers as you can.

   b   Which of the dimensions you've found would be impractical for boxes? Explain why.

   c   If you were Mr Johnson, what dimensions would you choose for the boxes? Why?

What did you learn?

1   How would you work out the volume of this mathematics book? Write down the steps you would follow.

2   Which has the greater volume: a brick 12 cm wide, 8 cm high and 23 cm long or a sponge 13 cm high, 10 cm wide and 15 cm long?

Topic 11 Review

Key ideas and concepts

Copy these sketches, label them and write short notes to summarise what you learnt in this topic.

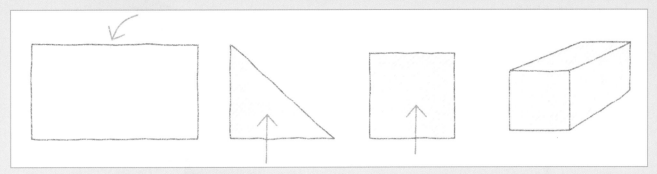

Think, talk, write ...

Look at your school shoe.

1   Estimate and measure the perimeter of your shoe print in millimetres.
2   How could you use squared paper to estimate the area of your shoe print?
3   What is the smallest size box that your shoe could fit into?
4   Draw a sketch and label it to show the dimensions.
5   Tell your group how you worked out the dimensions.

Quick check

1   Work out the perimeter and area of each of these shapes.

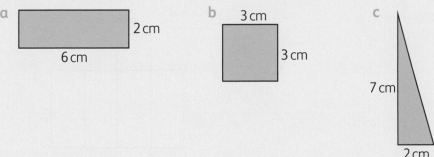

2   Explain in your own words:
    a   the main difference between calculating the area of a square and the area of a rectangle
    b   the relationship between the area of right-angled triangles and rectangles.

3   This is a diagram of a plot of land drawn at a scale of 1 : 200.
    a   Measure the lengths of the sides on the plan.
    b   Work out the real lengths using the scale.
    c   Calculate the perimeter of the plot of land.
    d   What is the area of the plot of land?
    e   What is the volume of a tank on the land that is 2 m long, 3 m high and $1\frac{1}{2}$ m wide?

1 : 200

123

Teaching notes

3-D objects and their parts

* Solid objects are three-dimensional (3-D) shapes because they have three dimensions that you can measure: length, breadth and height.
* Flat surfaces on 3-D objects are called faces.
* Some 3-D objects don't have faces but curved surfaces instead. Ball shapes, for example, have only one curved surface, no faces.
* Where two faces meet they form an edge.
* Any corner that is formed where three (or more) faces of a 3-D object meet is called a vertex (plural: vertices). The point of a cone is also called its apex.

Naming 3-D objects

* Solids are named according to the number and shape of their faces.
* A cube has six square faces.
* A cuboid or rectangular prism has six rectangular faces (some may be square).
* A cone has a flat circular base and a curved surface that forms a point (the apex).
* A cylinder has two flat circular end faces and a curved surface.
* A sphere (or ball shape) has one curved surface.

Nets

* A net is the template (or pattern) that you can fold up to make a model of a solid.
* All of the faces of the solid can be seen on the net. The fold lines on the net show where faces meet at an edge on the solid.

Coordinate systems

* In mathematics, the Cartesian plane is a special grid with an x axis and a y axis that is used to plot points and draw graphs.
* The position of any point on the grid can be given using two coordinates. These are given as a pair, and the x coordinate is always given first (x, y).
* To find the position of a given point, locate the numbers on the x axis and y axis and move up and across to the point where they meet.

A

Why are boxes like these called three-dimensional or 3-D objects? What does that mean? What do you call the flat surfaces of the boxes? Do any of them have curved surfaces? Which of these boxes do not have any vertices? Which box has the greatest volume?

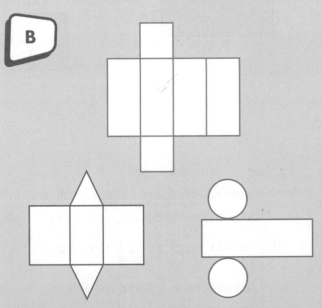

B

A designer has made and cut out these three pieces of card. How can she use these to make 3-D models? What 3-D object will she be able to make with each of them? Can you think of one thing in real life that is similar to each 3-D object?

C

y axis

x axis

How can you describe the position of the black and white horses on the chess board? What do you call a pair of numbers that describes position on a grid? Why do you think mathematicians have agreed that the *x* coordinate is always given first?

D

Look at the packaging of each item in the photograph. How do you think the designers worked out how much material is needed to make the packaging for each item? Which shapes are easy to pack into a box? Which are most difficult to display on a shelf? Why?

Think, talk and write

A   3-D objects and their properties *(pages 126–128)*

Read the descriptions and say which solid object each one could be.

1   I have six faces each with four edges and four vertices.

2   I have one circular face and one apex.

3   I have no faces, no vertices and no edges, but I am a 3-D object.

4   All my faces have only right angles.

B   Construct and draw 3-D objects *(pages 129–130)*

1   Choose one of these 3-D objects and try to draw it yourself.

2   What would the net of each shape look like? Draw a rough sketch of each one.

C   Coordinate systems *(pages 131–133)*

Each letter of the alphabet is located on this grid.

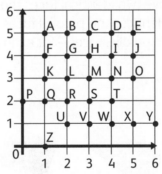

1   Write the coordinates of each letter in your first name.

2   Choose a word and write the coordinates of each letter. Ask a partner to work out what word it is.

3   What is the difference between the point (3, 4) and the point (4, 3) on a grid like this?

D   Shape and space at work *(pages 134–135)*

Cans of food are usually packed into boxes containing 24 cans.

1   How many ways can you find to arrange 24 cans to fit into a cuboid-shaped box?

2   Which arrangement uses the smallest box?

A 3-D objects and their properties

This is a **cube**. It has 6 square **faces**, 8 **vertices** where faces meet at a point and 12 **edges** between faces.

3-D objects or solids can be described and named according to their properties. For example:

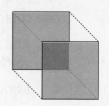

A **sphere** is a ball shape with no faces, vertices or edges.	A **cone** has a flat circular face and a curved surface that forms a point (called an **apex**).	A **cylinder** has two circular faces and one curved surface. It has no edges or vertices.
	apex face	 face face

Maths ideas

In this unit you will:
* use the properties of different 3-D objects to identify and name them
* identify shapes that are useful for different purposes in real life.

Key words

cube	apex
faces	cylinder
vertices	cuboids
edges	prisms
sphere	pyramids
cone	base

Cubes, **cuboids** and cylinders have two identical faces that are parallel to each other. These shapes can also be called **prisms**. The identical and parallel faces of a prism can be any shape. Prisms can be named using the shape of the parallel faces – for example, a cuboid could also be a square prism or a rectangular prism.

These are all prisms:

Cube

Cuboid

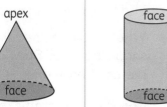
Cylinder Hexagonal prism Triangular prism Pentagonal prism

Pyramids are 3-D objects with a **base** (a bottom face) and triangular faces that meet at an apex. The base of the pyramid can be any polygon. Pyramids are sometimes named using the shape of their base. For example, a square-based pyramid has a square base and four triangular faces that meet at the apex.

Here are some different types of pyramids.

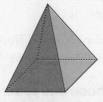
Square-based pyramid

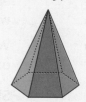
Hexagonal-based pyramid

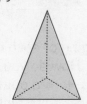
Triangular-based pyramid

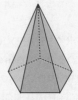
Pentagon-based pyramid

1   Name each solid as accurately as possible and list its properties.

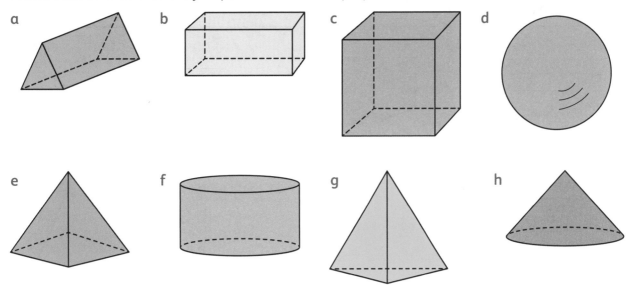

a   b   c   d

e   f   g   h

2   Look at the objects in Question 1 again.

a   Which three are most common in your environment? Give examples of where you might find them.

b   Which are least common?

c   Why do you think some solids are more commonly used in real life than others?

3   The items we use in our daily lives are solid 3-D objects.

a   Name the different types of 3-D objects you can see in this picture.

b   Could any of these items be made using a different 3-D object? Explain your answer and draw sketches to show what shape or shapes could be used.

4   Write down the letter of the object that is the odd one out in each group. Give a reason why it is the odd one out.

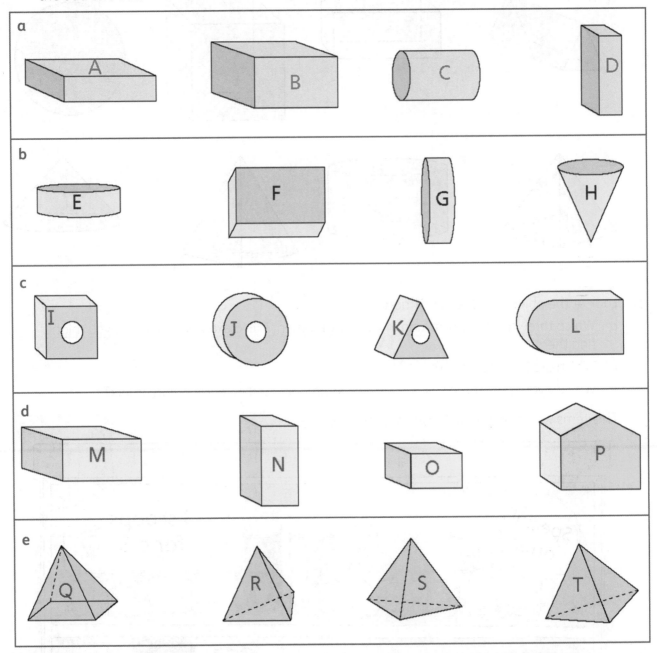

What did you learn?

Which 3-D object has:

1   a round base and a curved surface that comes to a point?

2   a pentagonal base and five triangular faces?

3   two pentagonal faces and five rectangular faces?

4   seven faces in total?

5   no faces and no edges?

6   an apex and six triangular faces?

B Construct and draw 3-D objects

Explain

You already know that a **net** is a pattern that shows the faces of a 3-D object and how they are connected to each other. Nets can be folded up to make **models** of 3-D objects. You can use the net on the left to build the cuboid on the right.

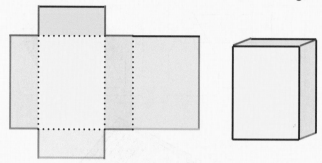

Maths ideas

In this unit you will:
* revise nets of 3-D objects and use them to construct models
* interpret drawings of 3-D objects
* draw 3-D objects.

Key words

net

models

1   Which of these nets could be used to make models of 3-D objects? You can redraw the nets on paper and fold them up to check whether you are correct.

a    b   c

d 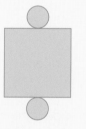   e 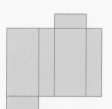   f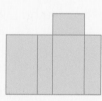

g 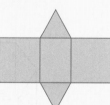   h 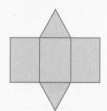   i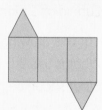

2   If you folded up this net, what would you end up with? Why?

3   Two faces are missing from this net. Copy and add the two missing faces and say what 3-D object you would make if you folded up the completed net.

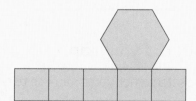

Explain

You can draw pictures of prisms starting with one of the parallel faces.
Follow these steps:

Step 1: Draw one of the parallel faces.	**Step 2:** Duplicate it.	**Step 3:** Draw lines to join matching vertices.	**Step 4:** Shade faces for effect.
	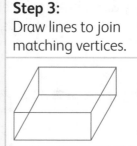		

Faces and angles look different on 3-D drawings.
Look at this drawing of a house.

All of the marked angles are right angles in real life. But if you measure them on the drawing, they are not right angles. The front door is a rectangle, but in the diagram it looks like a parallelogram because of the way you are viewing the house.

4   Use the method above to draw 3-D diagrams of:
    a a cube   **b** a triangular prism   **c** a cylinder.

5   Draw the shape you would make by combining these 3-D objects.
    a Cuboid and cube   **b** Cone and cone   **c** Cylinder and cone
    d Cuboid and triangular prism   **e** Cube and square-based pyramid

6   Look at this drawing of a plant pot.
    a What is this 3-D object called? Why?
    b What shape are the upright sides of the plant pot in real life?
    c Why do they not look like this shape on the drawing?
    d The base and the top of the plant pot are identical. Why does the base look different on the drawing?

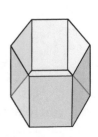

What did you learn?

1   Sketch a net you could use to build an open-topped cube.

2   Draw a sketch of a prism with the two parallel faces on the right.

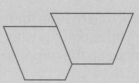

C Coordinate systems

You can use a pair of numbers to give the **position** of points on a **grid**.

The grid is formed by two number lines, called **axes**.

The *x axis* and *y axis* start from 0.
The point (0, 0) is called the **origin**.

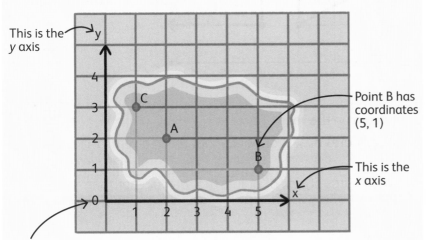

This is the *y* axis

Point B has coordinates (5, 1)

This is the *x* axis

This is the origin. It has the coordinates (0, 0)

Maths ideas

In this unit you will:
* revise how to use coordinates and a grid to give positions
* plot points on a grid and use coordinates to find points on a grid.

Key words

position	origin
grid	*x* coordinate
axes	*y* coordinate
x axis	coordinates
y axis	

Points A, B and C are shown on the grid.

Each point has an *x* **coordinate** and a *y* **coordinate**.

The *x* coordinate is always given first.

If the **coordinates** are (1, 3), this means that the point is level with 1 on the *x* axis and level with 3 on the *y* axis. This is point C on this grid.

1 The red points on the grid of this map mark the position of different places.

   a Give the position of:

   i the beach

   ii Rocky Ridge

   iii the airport.

   b Where would you be if you were at each of these positions?

   i (3, 2)

   ii (0, 4)

   iii (3, 4)

   c The Mangrove Reserve is marked at (6, 2). Give two other positions in the reserve.

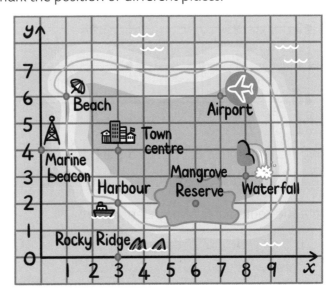

2 Start at the origin. Follow the route on the grid.

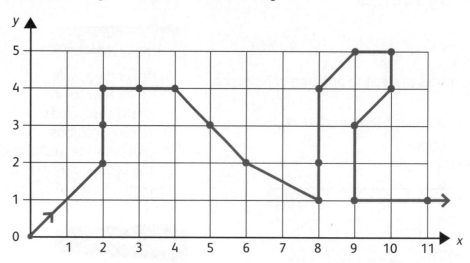

a Write the position of each dot marked on the route.

b What do you notice about the coordinates of sets of points that are in a vertical line? Can you explain this?

c What do you notice about the coordinates of sets of points that are in a horizontal line? Can you explain this?

3 Use the points on the grid.

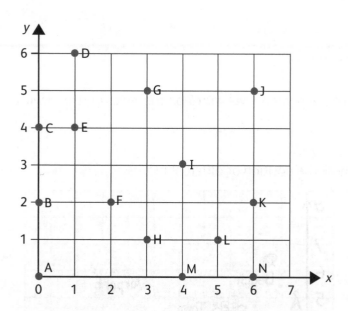

a Which letter is found at each of these positions?

   i (0, 4)   ii (4, 0)   iii (3, 1)   iv (2, 2)

b Write the coordinates of each letter.

   i B   ii L   iii A   iv C   v K

4   Mr Jessop likes to write coded notes to his students. This is the code board they use.

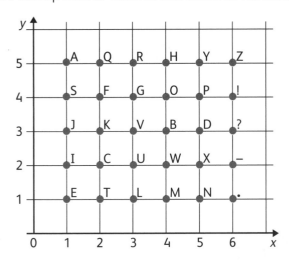

a   Decode this message:

(1, 4) (2, 1) (3, 2) (5, 3), (5, 5)

(4, 5) (1, 5) (3, 5) (5, 3) (1, 1) (3, 5) (6, 4)

b   Use the code board to write a short message of your own.

c   Swap your message with a partner and work out each other's messages.

What did you learn?

The map shows the route a fishing charter boat followed to get to the best fishing spot.

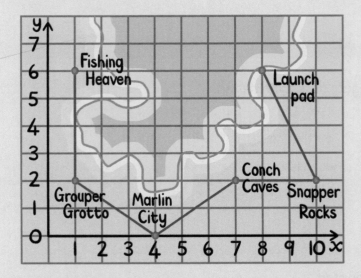

1   What are the coordinates of the launch pad?

2   Where is the boat when it is at (4, 0)?

3   What are the coordinates of Fishing Heaven?

4   Give the name and the coordinates of the first spot the boat reaches after Marlin City.

133

D Shape and space at work

Explain

Product designers and engineers use mathematics to help them design packaging and many other products. Measuring skills and a good understanding of 3-D objects, nets and scale models are useful for this work.

Maths ideas

In this unit you will:
* investigate how geometric ideas are used in daily life.

1   Look at these packaging shapes.

Choose two different packaging shapes from the photo and answer these questions about each one.

a   What does the packaging hold?

b   Describe each shape mathematically and sketch a rough net of it.

c   Design a differently shaped package for each product. Draw it and write two reasons why you think it is a suitable shape.

Investigate

2   You are going to make models and investigate packaging design.

You will need:

* a ruler
* a pair of compasses
* thick cardboard
* newspaper

* thin card
* scissors
* glue or tape.

Work with a partner.

Read through all the information first before you start.

The problem

Sharleen makes cookies and sells them to raise money for charity.

Each cookie is round and about 1 cm thick. The cookies each have a diameter of 8 cm.

8 cm

Sharleen wants to wrap 10 cookies in a clear plastic cylinder like this one.

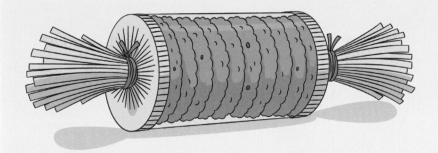

Sharleen's daughter draws this diagram to show her how to make the cylinder and how to work out the dimensions.

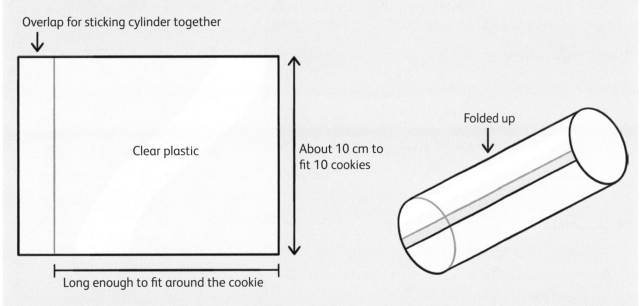

3   Make a model cookie. Draw a circle with a diameter of 8 cm. Cut out two or more circles of cardboard and stick them together to make a model about 1 cm thick.

4   Make a thin cardboard cylinder that will fit around the cookie and that will be long enough to hold ten cookies.

5   Work with newspaper and card to model lids for each end of the cylinder.

6   Calculate how much plastic you'll need for each cylinder.

7   How much material will you need for each lid?

8   Can you think of any other way of packing ten cookies? Sketch your ideas and show them to the class. Explain why you think this is a good way of packaging them.

What did you learn?

List three ways in which people use geometry in their daily lives or work.

Topic 12 Review

Key ideas and concepts

1   Make an illustrated poster to summarise what you learnt about 3-D objects in this topic. Your poster must include these headings:

   *   What is a 3-D object?

   *   Types of 3-D objects

   *   How to make models of 3-D objects using nets

   *   Drawing 3-D objects

   *   Using 3-D objects and their properties in real life

2   Draw a coordinate grid with at least five points on each axis and write notes to explain how you can use the grid to give the position of different points.

Think, talk, write ...

Answer these questions in your maths journal.

1   What did you find most interesting in this topic? Why?

2   What section was easiest for you to understand? Why?

3   Is there anything that you don't understand completely? If so, what can you do to learn more about it?

Quick check

1   Which 3-D objects have:
   a   no edges or vertices?
   b   round faces?
   c   curved surfaces?
   d   six identical faces?
   e   two parallel and identical pentagonal faces?

2   Which of these nets fold up to form a cuboid? Explain why the others won't work.

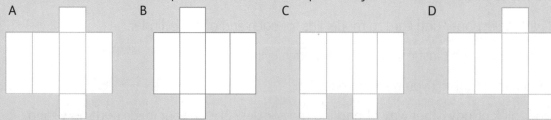

3   Draw a copy of each of these 3-D objects. Label your drawing with the name of the shape.

A

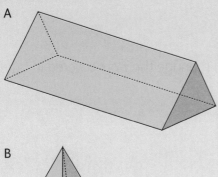

B

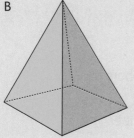

4   Look at this grid.

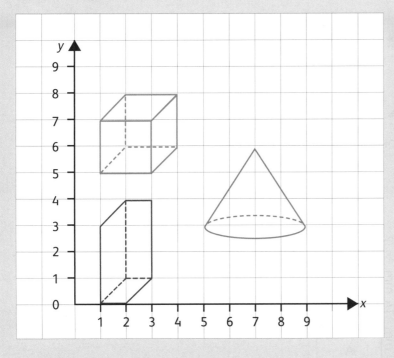

a   What do we call the point located at (0, 0)?

b   Write the coordinates of each vertex of the cuboid.

c   What are the coordinates of the apex of the cone?

d   What do you call the part of a 3-D object located at (2, 8)?

137

Test yourself (2)

Explain

Complete this test to check that you have understood and can manage the work covered in Topics 1 to 12.
Revise any sections that you find difficult.

1   From the set of numbers in the box, write down:

> 2, 4, 17, 24, 10, 41, 49, 12

   a   a multiple of 8   b   two square numbers
   c   a prime number   d   a factor of 8
   e   the lowest common multiple of 6 and 4   f   a triangular number.

2   Each block on this diagram represents a square with sides of 1 cm. What area does the drawing cover?

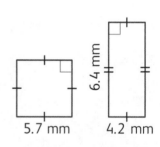

3   Write down the next three numbers in each of these number patterns.
   a   3, 6, 9, 12, 15, …
   b   10 000, 1 000, 100, …
   c   0, 1, 3, 6, 10, …

4   Calculate the area of each of these shapes.

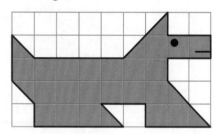

5.7 mm   4.2 mm   20 mm

5   This drawing of an ant has been drawn at a scale of 5 : 1.
   a   What does a scale of 5 : 1 mean?
   b   Is the ant in the picture bigger or smaller than a real ant?
   c   Measure the length of the ant on the drawing.
   d   Use the scale to work out how long the real ant would be.

6   Write each of these lengths in centimetres.
    a   68 mm   b   108 mm   c   3 metres   d   a kilometre

7   How many grams are there in each of these amounts?
    a   5 kg   b   130 mg   c   5.5 kg   d   8 700 mg

8   Convert each measurement to the units given.
    a   4.234 litres to millilitres   b   7.34 litres to centilitres   c   145 millilitres to litres

9   A teacher collected data about how many brothers and sisters students had. This is her data:

1 4 0 1 3   6 2 1 1 1   2 3 2 1 1   2 3 5 1 1   2 2 2 1 2

    a   What method do you think the teacher used to collect the data? Why?
    b   Draw a frequency table to organise the results.
    c   What is the mode of the data?

10   Find the mean and mode of these sets of data. Give your answers correct to one decimal place.
    a   2, 4, 2, 7, 3, 5, 4, 2, 3, 1
    b   40, 20, 30, 60, 50, 10

11   Write these numbers as products of their prime factors.
    a   18   b   48   c   100

12   Find the HCF of these numbers.
    a   28, 42   b   16, 20, 32

13   Write the name of the shape that best describes these objects.
    a   A box of tissues
    b   A tin of beans
    c   A dice
    d   Draw a rough sketch of the net that could be used to build each shape in parts a, b and c.
    e   Make a 3-D sketch of a tissue box.

14   This object was made by joining two cuboids. Calculate the volume of the whole shape.

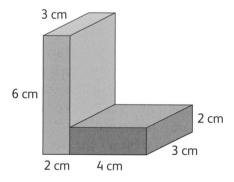

15 a   A carton of soap powder holds two boxes along its height and four boxes along its width. There are 48 boxes in the carton. How many boxes fit along the length of the carton?
    b   A toy box holds 36 building bricks. The bricks are packed in four layers, with three bricks along its length. How many bricks fit along the width of the box?

Teaching notes

Calculating with percentages

* Percentages can be written as fractions or decimals to make calculations easier. Make sure students remember how to convert between the two.

* To express one number as a percentage of another, you write the two amounts as a fraction and find the equivalent fraction with a denominator of 100, for example, $\frac{12}{20} \times \frac{5}{5} = \frac{60}{100} = 60\%$. You can also work this out as $\frac{12}{20} \times \frac{100}{1} = \frac{1\,200}{20} = 60\%$. On a calculator you would enter $12 \div 20 \times 100$ to get the same result.

Profit and loss

* Profit or loss is the difference between what you pay for an item (the cost) and what you sell it for. Actual profit or loss can be worked out by adding or subtracting decimals, and the answer is a money amount.

* Percentage profit or loss can be calculated using the formula
Profit (or loss) = $\frac{\text{cost price}}{\text{selling price}} \times 100$
Mathematically, this is the same calculation as expressing one number as a percentage of another.

Calculation with ratio

* Students need to be able to share quantities in a given ratio. To do this, they need to remember that a ratio of 2 : 3, for example, means 2 out of 5 parts and 3 out of 5 parts and not $\frac{2}{3}$.

* Students also need to remember how to calculate and manipulate equivalent ratios in order to solve problems involving ratio.

Project work

* Projects involve studying a particular topic in some detail (and projects may go across subject areas).

* When students do projects, they are required to do their own investigations or research. Once they have done this, they need to report their findings. They can do this by writing a report, making a poster, or doing an oral and/or audio-visual presentation.

* The project in this topic takes students through some of the skills they need to use in their projects. Adapt this project to your classroom needs if you wish to.

A

Sue makes small pottery dishes to sell at a school market. How can she work out what it costs to make each dish? She wants to sell the dishes and make an average profit of 15%. How can she work out at what price to sell them? Should she make all the dishes the same price? Give reasons for your opinion.

Think, talk and write

A Calculating with percentages
(pages 142–146)

1 A school tuck shop orders 400 packets of nuts. They sell 22% of them on Monday.
  - a What percentage of the packets is left?
  - b How many packets were sold?
  - c If the tuck shop sells 50% of the remaining packets on Tuesday, how many packets are still in stock?

2 How much will James save if he buys a $500 bicycle and gets a discount of 15%?

3 Explain in your own words what these terms mean.
  - a Profit
  - b Loss
  - c Selling price
  - d Discount

B Calculating with ratios *(pages 147–148)*

1 Look at the chocolates in the photo again. How many would each person get if the chocolates were shared in the ratio:
  - a 1 : 1?
  - b 1 : 2?
  - c 1 : 2 : 3?
  - d 12 : 4?

2 To make pink paint, a hardware store mixes red and white paint in the ratio 1 : 4.
  - a What fraction of a tin of pink paint will be red paint?
  - b If the shop uses 200 ml of red paint, how much white paint will they need?
  - c If there are 480 ml of white paint in the mixture, how much red paint will there be?

3 Discuss how you could make the paint a paler shade of pink. Suggest a ratio of red : white that would result in a paler pink shade. How could you make a darker pink?

B

Look at this tray of chocolates. What is the ratio of cuboid-shaped chocolates to other shapes? What is the equivalent ratio in the form of 1 : ☐? If two brothers share the chocolates in the ratio 3 : 5, how many will each brother get?

A Calculating with percentages

Explain

Read through the information and examples to remind yourself how to calculate with **percentages**. To write one amount as a percentage of another, you form a fraction and convert it to a percentage.

Example 1

There are 450 students in a school, and 216 of them are girls. What percentage of the students are girls?

216 out of 450 students are girls, so:

$\frac{216}{450} \times 100 = 48\%$

To calculate a percentage of an amount, write the percentage as a fraction with a denominator of 100 and find this fraction of the amount (remember 'of' means multiply). You can cancel to make it easier to work with the numbers.

Example 2

What is 40% of 320?

$\frac{40}{100} \times \frac{320}{1} = \frac{40}{100} \times \frac{320}{1} = 4 \times 32 = 128$

You can work with percentages greater than 100% in the same way.

Example 3

What is 110% of 600?

$\frac{110}{100} \times \frac{600}{1} = \frac{66\,000}{100} = 660$

You can calculate the total amount if you know a percentage.

Example 4

If 15% of an amount is 60, what is the whole amount?

15 **per cent** = 60, so 1% = $\frac{60}{15}$ = 4

The whole amount is 100%, so multiply 4 (which is 1%) by 100 to work this out.

4 × 100 = 400

This is the same calculation as $\frac{60}{15} \times 100$.

Maths ideas

In this unit you will:
* calculate percentages of quantities
* revisit profit and loss and how to calculate these
* express profit or loss as a percentage
* solve problems involving percentages.

Key words

percentages	profit
per cent	loss
increase	cost price
decrease	selling price
discount	

1   Calculate.

    a 2% of 200   **b** 10% of 600   **c** 50% of 200   **d** 20% of 80

    e 75% of 50   **f** 5% of 40   **g** 100% of 72   **h** 15% of 60

2   Calculate, giving the answers as decimals if necessary.

    a 10% of 85   **b** 5% of 171   **c** 15% of 90

    d 20% of 16   **e** 15% of 300   **f** 78% of 144

    g 110% of 500   **h** 125% of 300   **i** 105% of 206

3   In a school of 400 students, 30% are boys.
    a   How many boys are there?
    b   How many girls are there?

4   A fish weighed 6.4 kg. A fishmonger dried it, and it lost 25% of its weight. How many kilograms did the fish lose through drying?

5   Work out the value of *n* in each of these statements.
    a   15 is *n*% of 30
    b   40 is *n*% of 80
    c   15 is *n*% of 45
    d   *n*% of 50 = 10
    e   *n*% of 80 = 36
    f   *n*% of 120 = 60

6   a   What percentage of 50 is 20?
    b   What percentage of 150 is 90?

7   After travelling 40 km, a motorist still has 60% of the journey to travel. What is the total length of the journey?

8   Answer these questions. Show how you worked out the answers.
    a   50% of a number is 25. What is the number?
    b   45% of a number is 45. What is the number?
    c   80% of John's salary is $48. What is his salary?

Problem-solving

9   Jeremy scored 60% of the total marks in a test. If the test was marked out of 75, what was his score?

10 Sammy was given 75% of a sum of money. If Sammy received $120, what was the total sum of money?

Explain

You can **increase** or **decrease** amounts by percentages.

Example 1

4 460 tickets were sold for the first match in an international cricket tournament. Sales increased by 5% for the second match. How many tickets were sold for the second match?

The increase is 5% of 4 460.

$\frac{5}{100} \times 4\,460 = 223$ So 223 more tickets were sold for the second match.

The total sales for the second match were 4 460 + 223 = 4 683

Example 2

A DJ usually charges $3 500 to play at a festival, but he agrees to give the organisers a 15% **discount**. How much will they have to pay?

Discount = 15% of $3 500

$\frac{15}{100} \times 3\ 500 = 15 \times 35 = \$525$

They will pay $3 500 – $525 = $2 975

You also use percentage increase and decrease when you work with profit and loss.

---

11 Calculate the sale price of each pair of shoes.

12 The cost of wood went up and a carpenter had to put her prices up by 15%. The picture shows the old prices of the furniture.

Work out the new price of:

a a large table

b a set of chairs

c a small table.

Explain

Percentages, profit and loss

If you buy or make something and sell it for more than it cost you, you make a **profit**. If you have to sell it for less than it cost you, you make a **loss**.

Cost price + profit = selling price
Cost price − loss = selling price

You can calculate the percentage profit or loss if you know the **cost price** and the **selling price**.

Example

Mario bought a machine for $14 000 and sold it three years later for $12 600.

Calculate the percentage profit or loss he made on the deal.

$12 600 < $14 000

He sold it for less than he paid, so he made a loss.

The price difference is $14 000 − $12 600 = $1 400

Next, make a fraction using the difference and the original price and convert it to a percentage.

$\frac{1\ 400}{14\ 000} \times 100 = 10\%$ loss

13 Calculate the amount of profit or loss in dollars.
    a 25% profit on $300
    b 10% loss on $520
    c 25% loss on $12 000

14 Nadia bought a second-hand phone for $480. She sold it online and made 10% profit. What price did she sell it for?

15 A farmer bought equipment for $5 000, but then found that it wasn't right for him, so he sold it for $4 200. Calculate his percentage loss.

16 A vendor bought 100 T-shirts to sell at the market and paid $400 for them. She sold them for $8 each.
    a Did she make a profit or a loss?
    b Calculate the percentage profit or loss she made.

What did you learn?

1 5% of a number is 15. Work out:
    a 50% of the number
    b 98% of the number
    c 2% of the number.

2 A shop advertises a table for $749 and a bookshelf for $550. The items get dirty, so the shopkeeper decreases the prices by 20%. What are the new prices?

3 Use the price information below and work out the percentage profit or loss on each sale.
    a Cost price $700, sold for $630
    b Sold for $7 200, cost $6 000 to make

Project

Working out costs and prices

You are going to work in groups to make craft items that you can sell at a fête to raise money for charity. You need to decide what you will make, what you will need to make it and how much it will cost to make one item. Then you will need to decide what price to sell the items for so that you can make a profit to give to the charity. Use the planning sheets on pages 72 to 74 of Workbook 6 to record your ideas.

1   Start by brainstorming ideas of things you can make. You can write your ideas and sketch them, or you can find pictures to show what you want to make. Here are some ideas to get you started.

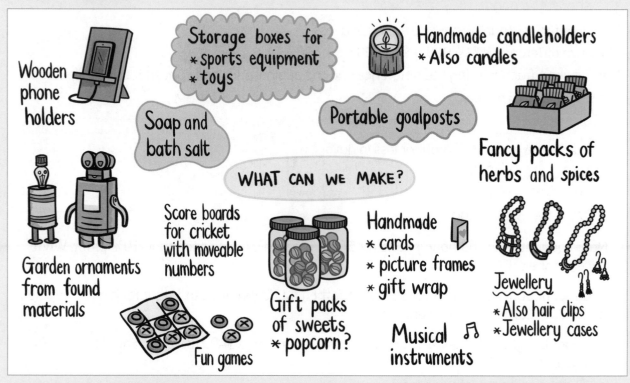

2   Decide what you will make.
    a   What materials will you need?   b   Where will you get them?

3   Allocate tasks to group members and make a prototype of the item you've decided to make.

4   Use the prototype to work out the cost of making one item. Costs will include materials you need to buy, transport to and from shops, packaging, labels and any other items that cost money.

5   Will you save on costs if you make 25 of the same thing? Work out the cost of making 25 items.

6   Decide what a good selling price will be for the item. You will need to cover your costs (if there are any) and also make a profit so that you can donate the profit to charity.

7   Present your ideas and your prototype to the class. Discuss whether the pricing you have chosen is reasonable or not.

B Calculating with ratios

Explain

You already know how to find **equivalent ratios** and how to write ratios in **simplest form**. You can use equivalent ratios to share amounts in a given ratio.

Example 1

Torianne and Maria share the cost of a $5 raffle ticket in the ratio 2 : 3. They agree to share any winnings in the same ratio.

How would they share a prize of $50?

Every share requires 2 parts for Torianne and 3 parts for Maria (five parts per share).

$\frac{2}{5}$ of 50 = $\frac{2}{5} \times \frac{50}{1} = \frac{100}{5}$ = $20

$\frac{3}{5}$ of 50 = $\frac{3}{5} \times \frac{50}{1} = \frac{150}{5}$ = $30

Once you've worked out Torianne's share, you can also subtract it from the total to find Maria's share.

Example 2

An artist mixes 1 part blue paint to 3 parts red paint to make purple paint. If he uses 200 mℓ of blue paint, how much purple paint will he make?

Ratio of blue to red is 1 : 3

So, 3 ml of the red paint are needed for every 1 mℓ of blue paint.

$$
\begin{array}{ccc}
1 & : & 3 \\
\times 200 & & \times 200 \\
200 & : & 600
\end{array}
$$

200 times as much red is needed for 200 ml of blue

200 + 600 = 800 mℓ

Add these amounts to find the total

1   Martin, John and Troy are eating mangoes. For every one that Martin eats, John eats two and Troy eats four.

    a   How many would they each have if they ate 21 altogether?

    b   How many do John and Troy eat if Martin eats four?

    c   If John eats 6, how many do the three boys eat altogether?

2   A person earns $2 300. Of each dollar, he spends 30¢ on food, 12¢ on clothing and toiletries and 11¢ on electricity and services.

    a   How much of the $2 300 does he spend on food?

    b   How much money does he spend on clothing and toiletries?

    c   How much money is left after he has paid for food, clothing and toiletries, and electricity and services?

Problem-solving

3   Look at the diagram carefully. Work out the length of the toy truck in paperclips.

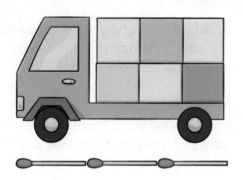

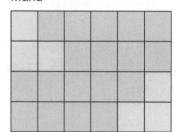

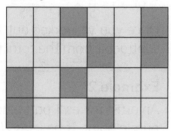

4   Look at these three flag designs made by different students.

Maria

Robert

Glenroy

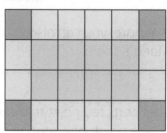

a   Write the ratio of yellow to other colours for each flag.

b   If Maria used 48 cm² of fabric to make her flag, how much blue material would she need?

c   Robert made a flag that is 9 cm long and 4 cm wide. Work out the total area of fabric he used and the area of each colour.

d   Glenroy made a small flag that is 12 cm². Work out the area of each colour on his flag.

What did you learn?

1   Share $40 in these ratios.
   a   1 : 3
   b   2 : 3
   c   6 : 4
   d   4 : 1

2   Two people share a prize in the same ratio as they paid for the ticket. If one share of the $55 prize is $22, what was the sharing ratio?

Topic 13 Review

Key ideas and concepts

Write short notes to answer these questions and summarise what you learnt in this topic.

1   How do you find a percentage of an amount?

2   How do you find an amount if you know 12% of the amount is 60?

3   What do the terms profit and loss mean?

4   How do you find the price you will pay if you will get a 20% discount?

5   If you add $5 to a cost price of $80, how do you work out the percentage profit?

6   What does it mean if you share an amount in the ratio 3 : 7?

7   If there are 300 people at a concert and 120 of them are female, how do you work out the ratio of female to male concert-goers?

Think, talk, write ...

1   a   A part-time worker in a shop is paid $25 per hour. She is offered the choice between an increase of 20% per hour or an increase of $7 per hour. Which option should she take? Why?

    b   A different worker is paid $40 per hour. She is given the same choice (20% or $7 per hour). Which offer is better for her? Why?

2   If you share things equally, what is the ratio of one share to the other? How does the ratio change if one person gets three times as much as the other?

Quick check

1   What is:

    a   15% of $300?   b   8% of 400 people?

    c   75% of 60 oranges?   d   60% of $40?

    e   35% of 150 mangoes?   f   12% of $500?

2   In a school of 900 students, 46% of students walk to school and 35% travel by bus. The other students travel by other means.

    a   What percentage of students travel by other means?

    b   Work out how many students use each mode of transport to school.

    c   The ratio of boys to girls in the school is 4 : 5. Calculate how many boys there are.

3   A Parish Council budgeted $500 000 for refuse removal. At the end of the year they worked out that they had spent 135% of the amount budgeted. How much did they spend on refuse removal?

4   A computer was sold for $3 000. The seller made a profit of 20%. What was the cost price of the computer?

5   A pack of sweets contains green and yellow sweets in the ratio 5 : 2.

    a   What does a ratio of 5 : 2 mean?

    b   If there are 42 sweets in the pack, how many are yellow?

    c   If there are 24 yellow sweets in the pack, how many green ones will there be?

Topic 14 Measurement (3)

Teaching notes

Time

* Students should be able to use the 12-hour and 24-hour clock confidently by now.
* You may need to remind students that the duration of an event is how long it takes. An event that starts at 05:00 and ends at 05:30 has a duration of 30 minutes.
* Counting on by adding first the hours and then the minutes is a useful way of working out duration.
* Time is not metric, so students need to remember the relationships between units when they calculate with time. Half an hour can be written as 0.5 hours, but this means 30 minutes and not 50 minutes.

Time, distance and speed

* Speed tells you how fast something is moving. Mathematically, speed is a rate that compares distance travelled and time taken. A speed of 45 km per hour means that a vehicle will cover an average distance of 45 km every hour.
* Students need to understand the relationship between time, distance and speed. They should be able to use it to generate formulae, for example: speed = distance ÷ time.

Money

* Remind students that money calculations can be treated like any other decimal calculation, as long as they understand that money amounts always have two decimal places, so we write $2.50 and not $2.5.
* Different countries have their own currencies, and these have different values, so when you convert from one currency to another, you need to know what the rate of exchange is. These rates change all the time. Online sites are useful for finding the latest rates. Show students how to use the automatic currency conversion tools that these sites offer.

Temperature

* Students should be familiar with the concept of temperature and how it is measured.
* The focus this year is on comparing the Celsius and Fahrenheit scales. Students need to work with real thermometers to estimate, compare and read temperatures.

A

This is a stopwatch. What is the difference between a stopwatch and a normal clock or watch? What do you measure using a stopwatch? What does the measurement show on this stopwatch? Suggest one event or activity that might have taken this amount of time.

B

What does this sign tell drivers to do? Why do you think we need signs like this one on the roads? What happens if drivers don't obey these signs? How do traffic authorities check whether a driver is obeying a sign like this or not?

C

Lynda's mom has travelled all over the world. She found these coins and notes at home. How do you know where coins and notes come from? Her mom says the coins are from Germany, Norway and the United Arab Emirates and the 5-peso note is from Argentina. Can you work out which coin is from which country?

D

Look at these two weather forecasts for Kingstown in St Vincent and the Grenadines. Both forecasts are for the same time period. How are they the same? How are they different? Which one shows degrees Celsius? How do you know this?

Think, talk and write

A   Working with time *(pages 152–153)*

1   Write each of these times in a different way.
   a   7:00 p.m.   b   08:00
   c   Four thirty in the afternoon
   d   Midday
   e   Half past two in the morning
   f   Five minutes after midnight

2   Terri's uncle starts work at 08:15 every day. He works until 16:45. He has an hour's break for lunch each day and he works five days a week. How many hours does he work in one week?

B   Time, distance and speed *(pages 154–155)*

1   If Mr Sands travels 168 km in two hours, what is his rate of travel in kilometres per hour? How did you work this out?

2   A runner takes 20 seconds to complete a lap of a track. How long will it take her to run these distances at the same speed?
   a   2 laps   b   6 laps   c   10 laps

3   Terry cycles 12 kilometres in an hour. How far will he have cycled in:
   a   2 hours?   b $3\frac{1}{2}$ hours?   c   0.5 hours?

C   Money *(pages 156–157)*

1   Calculate.
   a $32.65 + $23.08 – $12.00
   b $50 – $14.99

2   Debbie gets a bill for $28.75 and pays with two twenty-dollar bills. How much change will she receive?

D   Temperature *(page 158)*

1   What is temperature and how is it measured?

2   The temperature in a town in the Caribbean was 34 degrees Celsius at midday. By 6 o'clock in the evening the temperature had cooled down by 4 degrees and by 23:00 the temperature had cooled down by a further 3 degrees. What was the temperature at 11 o'clock that evening?

A Working with time

Maths ideas

In this unit you will:
* revise what you already know about time
* use the 12-hour and 24-hour system to tell and write times
* create and solve problems involving time.

Explain

You already know how to tell and write times using the 12-hour time system with a.m. and p.m. and the **24-hour time** system using four digits.
These clocks show the same time.

The time is 10 o'clock in the evening or 10 p.m.
You can also give this time as 22:00 (twenty-two hundred hours).

Key words

24-hour time
duration
elapsed

1   What is the time:
    a   35 minutes after 11:45 a.m.?
    b   half an hour before 12 noon?
    c   1 hour after midnight?
    d   3 hours after 10 a.m.?
    e   20 minutes after 6:45 p.m.?
    f   25 minutes after 11:35 a.m.?

2   Look at the flight information on the board.

Remember

* The **duration** of an event tells you how long something takes from start to finish.
* **Elapsed** time is the amount of time that passes from the start of an event until it finishes.

Flight number	From	Time due	Actual arrival
704	New York	6:05 p.m.	6:20 p.m.
501	Barbados	7:00 p.m.	7:00 p.m.
463	Trinidad	10:05 p.m.	9:45 p.m.

    a   Rewrite the times on the board using the 24-hour time system.
    b   Sue's aunt is arriving on Flight 704 from New York. How many minutes late is the flight?
    c   Is the flight from Trinidad late or early, and by how many minutes?
    d   A plane due at 16:20 was 50 minutes late. At what time did it arrive?
    e   An airplane leaves St Lucia at 15:50. The flying time to Martinique is 15 minutes. At what time is it expected to land?

3   Write these times using the 12-hour system. Remember to include a.m. or p.m.
   a   10:30
   b   13:45
   c   19:05
   d   23:49

4   Look at the airline timetable. It shows the times that four planes leave Toronto and arrive in Antigua.

Flight	Leaves Toronto	Arrives Antigua
CPM-1	10:05	16:30
CPM-2	14:15	21:50
CPM-3	15:30	23:20
CPM-4	16:45	22:40

   a   Which is the fastest flight?
   b   Which is the slowest flight?
   c   Flight CPM-2 leaves Toronto on time, but arrives 45 minutes late in Antigua owing to bad weather on the way. What is the duration of the flight?
   d   Sam arrives at the airport at 16:15 to fetch his sister. He waits for $1\frac{1}{4}$ hours for her to arrive. At what time does she arrive?

5   Work out when each activity finished.
   a   I go for a run at 13:55. I run for an hour and ten minutes.
   b   A TV show starts at 22:05. It lasts for 45 minutes.
   c   Sports day starts at 8:30 a.m. It finishes $4\frac{3}{4}$ hours later.

6   Work out the starting time of each activity.
   a   I get home at 18:10 after a 55-minute car journey. What time did I start the journey?
   b   A movie finished at 23:25. It was 2 hours 45 minutes long. What time did it start?

7   Work out how long each activity took.
   a   James put a cake in the oven at 13:07 and took it out at a quarter to two. How long did he bake the cake for?
   b   Rebecca started a car journey at 13:38 and arrived at her destination at 14:05. How long did the journey take?
   c   Jenna started her run at 18:27 and finished at 7:07 p.m. How long did she run for?

What did you learn?

1   A bus left a village at ten to two in the afternoon and arrived at Castries at 15:30. How long did the journey take?

2   The bus left again at quarter to four in the afternoon. Write this time in three different ways.

B Time, distance and speed

When we travel, we often need to work out the speed at which we are travelling. **Speed** is a **rate** that tells us how much time it takes to cover a given **distance**.

You can say that a car is driving at 60 kilometres (distance) per hour (time), or 60 km/h. This means, if the car continues at the same speed for one hour, it will cover 60 kilometres.

When athletes run a 100-m sprint, their speed is measured in metres per second.

Sometimes the speed of an object is given as a rate that you can work out quickly. An airplane that covers a distance of 860 km in one hour has an average speed of 860 km/h. A car that drives 50 km in half an hour travels at a speed of 100 km/h.

Some rates are more complicated to work out. For example, what is the speed in km/h of a cyclist who covers 23 km in 45 minutes?

Maths ideas

In this unit you will:
* learn about average speed and how to calculate it
* understand the relationship between time, distance and speed
* solve problems involving time, distance and speed.

Key words

speed	formula
rate	relationship
distance	

To calculate the speed of a travelling object, you can use the **formula** or rule:

Speed = Distance ÷ Time, or $S = \dfrac{D}{T}$

Example 1

A truck leaves the harbour at 4:10 p.m. and arrives at the warehouse at 5:40 p.m. The warehouse is 84 km from the harbour. Did the truck driver exceed an average speed of 60 km/h?

You can use the formula to calculate the truck's average speed. Remember that the time must be in hours and not in minutes.

Speed = $\dfrac{84 \text{ km}}{1.5 \text{ hours}}$

  = 56 km/h

The truck travelled at an average speed of 56 km/h and therefore didn't exceed 60 km/h.

You can also use this **relationship** between distance, time and speed to determine the distance travelled when you know the speed and the time:

Distance = Speed × Time   D = S × T

Example 2

Darius walked from his home to his grandparents' home at a steady speed of 5 km/h. He walked for 3 hours. How far did Darius walk?

Use the formula:

Distance = Speed × Time

  = 5 km/h × 3 h = 15 km

Darius walked 15 km in 3 hours.

To work out the time taken when you know the distance and speed, the formula looks like this:

Time = Distance ÷ Speed $T = \dfrac{D}{S}$

Example 3

Trudy cycles at 8 km/h and covers a distance of 20 km. How long does her journey take?

Time = $\dfrac{\text{Distance}}{\text{Speed}}$

$\dfrac{20}{8}$ Think of this as an improper fraction and simplify it

$= 2\dfrac{4}{8}$

$= 2\dfrac{1}{2}$ Her journey took $2\dfrac{1}{2}$ hours.

Remember

This triangle diagram shows the relationship between speed, distance and time. You can use it to work out the formula you need.

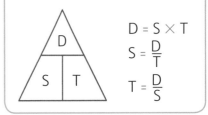

$D = S \times T$

$S = \dfrac{D}{T}$

$T = \dfrac{D}{S}$

1   Try to work out these answers mentally.

  a   If it takes you half an hour to walk three kilometres, how far will you go if you continue walking for an hour?

  b   A car covers a distance of 180 kilometres in two hours. What distance does it cover in one hour if it travels at the same speed throughout?

  c   Stefan rides his bike at a constant speed of 10 kilometres per hour. How long will it take him to travel a distance of 15 kilometres?

2   A car travels 240 km at an average speed of 50 km/h. How long does the trip take?

3   Mr Brown drove his car at an average speed of 80 km/h. How many kilometres did he travel in:

  a   4 hours?   b $5\dfrac{1}{2}$ hours?   c $2\dfrac{1}{4}$ hours?

4   Tyrone drove for 400 km at an average speed of 80 km/h. How long did he drive?

5   Bernadette runs from 10:40 a.m. until 11:10 a.m. at an average speed of 9 km/h. How far does she run?

Problem-solving

6   Perry ran 4 500 m in half an hour and Jerome ran 6 000 m in 40 minutes. Who ran faster?

7   What is the speed of a plane that travels 243 km in 2 h 15 min?

8   Melanie spent four hours on a plane. For half of that time, the plane's average speed was 900 km/h and for the rest of the time, its average speed was 760 km/h. What distance did Melanie travel?

What did you learn?

1   What would you mean if you said that you were travelling at 80 km/h?

2   How do you calculate speed, distance and time?

C Money

You worked with money amounts in Topic 13 when you dealt with profit, loss and discount.

When you write money amounts, remember that the decimal point separates the dollars from the cents. For example, $2.45 means 2 dollar and 45 cents.

Money amounts have two decimal places, so you write $45.50 and not $45.5.

You do not write the cent symbol (¢) in dollar amounts.

Maths ideas

In this unit you will:
* calculate with money amounts
* learn about foreign exchange rates
* convert between different currencies.

Key words

currency	foreign currency
convert	exchange rate

1   Write the price of each vintage car in words.

$215 600.00

$8 061 918.00

$1 000 754.00

2   Calculate.

a	$2.35 + $6.58	b	$16.39 + $12.47	c	$48.95 + $12.25
d	$173.69 + $89.35	e	$4.09 − $2.06	f	$7.20 − $5.85
g	$31.09 − $18.25	h	$129.35 − $106.27		

Investigate

3   Go to a clothing store, or look at an advertisement in a newspaper. Find prices for the following items:

T-shirt

Cap

Pair of shoes

Pair of jeans

Pair of sneakers

Shorts

Estimate, to the nearest dollar, the totals of these items.

a   Two pairs of sneakers and a T-shirt

b   Three pairs of jeans and four pairs of shoes

c   Six caps and a pair of shorts

d   Use your calculator to help you work out the actual total for each set of items.

e   List the notes and coins you could use to pay each total without getting change.

f   List the smallest number of notes you could use to pay each total, and say how much change you would get.

Explain

Different countries use different types of money. The money used in a country is called its **currency**. Some currencies used in the Caribbean are: East Caribbean dollars (EC$), Barbadian dollars (BDS$), Jamaican dollars (JA$) and United States dollars (US$).

When you travel to another country you may have to **convert** some of your local currency to **foreign currency**. The **exchange rate** for currencies is determined by banks and governments and can change from day to day.

You use the exchange rate and divide or multiply to convert from one currency to another.

Think and talk

Find out the names of some other currencies that are in use in Caribbean countries.

Example 1

The exchange rate for BDS$ to US$ is US$1.00 = BDS$2.00.
How many US dollars would you need to buy BDS$250.00?
250.00 ÷ 2.00 = 125.00
You would need US$125.00.

Example 2

The exchange rate for EC$ to US$ is EC$2.70 = US$1.00
How much EC currency do you need to buy US$150.00?
150.00 × 2.70 = 405.00
You would need EC$405.00.

---

4   At a rate of US$1.00 to BDS$2.00, how many US$ do you need to buy:
   a   BDS$101.00?   b   BDS$184.00?   c   BDS$349.38?

5   If the exchange rate is BDS$1.00 to EC$1.35, how much EC currency do you need to buy:
   a   BDS$5.00?   b   BDS$26.00?   c   BDS$45.00?

6   US$1.00 = EC$2.65. Copy these and complete them.
   a   US$7.00 = EC$_____
   b   US$120.00 = EC$_____

7   Find out the exchange rate for your country's currency to each of the following currencies.
   a   British pound   b   Euro   c   Australian dollar

What did you learn?

1   Josh pays $34.65 for a shirt. How much change will he get if he pays with a $50 bill?

2   If EC$1.00 = TT$2.50, convert the following Trinidad and Tobago dollars to EC dollars.
   a   TT$750.00   b   TT$381.00   c   TT$92.50

D Temperature

There are two scales of measurement for temperature: **degrees Celsius (°C)** and **degrees Fahrenheit (°F)**. The units on both scales are degrees.

This thermometer shows 0 °C, which is the same as 32 °F.

These are the rules for converting between degrees Celsius and degrees Fahrenheit:

* for converting °F to °C, subtract 32, multiply by 5, then divide by 9.
* for converting °C to °F, multiply by 9, divide by 5, then add 32.

Maths ideas

In this unit you will:
* read and compare temperatures using the Celsius and Fahrenheit scales.

Key words

degrees Celsius (°C)

degrees Fahrenheit (°F)

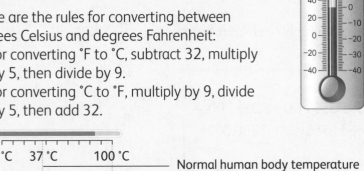
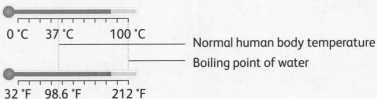

Normal human body temperature

Boiling point of water

1   Which temperature matches each situation the best?

   a   The temperature of water from a hot water tap

      42 °C   20 °C   36 °F

   b   The temperature of water from a cold water tap

      55 °C   22 °C   12 °F

   c   Normal body temperature

      36 °C   96 °C   55 °F

   d   The temperature in a freezer

      −15 °C   15 °C   150 °F

2   Sugar melts at 320 °F. What is this temperature in Celsius?

3   Record the temperature on each thermometer in degrees Celsius. Calculate how many degrees each temperature would need to rise to reach the boiling point of water.

a    b    c

What did you learn?

1   Give any two typical temperatures in °C and °F (for example, body temperature, or the freezing or boiling point of water).

2   Which temperature is colder: 0 °C or 0 °F?

Topic 14 Review

Key ideas and concepts

1   Copy and complete this diagram.

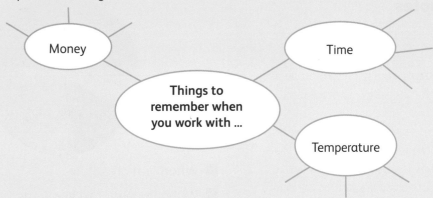

2   Draw the distance, time, speed pyramid and write short notes to explain how to work out each measurement (D, T and S).

Think, talk, write …

1   How can a change in the exchange rate cause the price of an overseas vacation to increase or decrease?

2   Why is it important to know what scale is being used when you work with different temperatures?

Quick check

1   Write the current time in two different ways.

2   Darren takes 12 minutes to walk to the bus stop, where he waits 9 minutes for the bus. The bus ride is 54 minutes long. How long does his journey take altogether? Give your answer in hours and minutes.

3   Nadia started reading at 09:15 and finished her book at 13:08. How long did she read?

4   What distance would you travel if you drove for 4 hours at 80 km/h?

5   How long does it take to cycle 35 km at an average speed of 7 km/h?

6   Mr Davis travelled 450 kilometres in 5 hours.
    a   What was his speed in km/h?
    b   How far would you expect him to travel in $7\frac{1}{2}$ hours if he kept to the same speed?
    c   How long would it take to cover 1 080 km at this speed?

7   Zara is from Trinidad. She buys EC$400 for a trip. How many TT$ is this if the exchange rate is TT$2.50 to EC$1.00?

8   Give these temperatures for both temperature scales.
    a   Healthy body temperature
    b   Boiling point of water
    c   Freezing point of water
    d   The temperature in your country in December

Teaching notes

Reading and interpreting graphs

* The heading, the labels and the scale (interval) on a graph are important. Students need to read these carefully to make sense of the data.
* Students have to understand that double bar graphs show two related sets of data on the same set of axes. Make sure they know how to compare sets of data represented in this way.
* Line graphs are useful for finding trends or patterns in a data set. The slope of a graph shows whether the values (such as temperatures) are increasing or decreasing.

Drawing graphs

* Students already know how to draw pie charts, bar graphs and line graphs. In this topic they will revise what they know and make choices about the best type of graph for different data sets.
* Remind them that all graphs need clear and precise labelling so that the reader can interpret the data correctly.

Probability

* The theoretical probability of something happening, for example, a coin landing on heads or tails, can be worked out mathematically. In the example of the coin, there are only two possible outcomes: heads or tails. This means there is an equal chance of heads or tails.
* The probability scale extends from 0 (impossible) to 1 (certain). The probability of different events occurring can be given as a fraction by comparing the number of favourable outcomes with the total number of outcomes. So the probability of getting a 1 when you roll a normal dice is 1 out of 6 possible outcomes, or $\frac{1}{6}$. The probability of rolling an even number is 3 out of 6 possible outcomes, or $\frac{1}{2}$.

A

When do you look at your mobile phone?

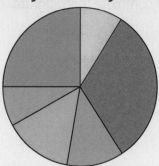

- ☐ When you are alone
- ☐ When you want to know the time
- ☐ When you are bored in class or at home
- ☐ When you do not feel like doing another task
- ☐ When you are eating a meal
- ☐ When your friends are looking at their phones

What does the pie chart show? How do you think the person who drew this pie chart collected and organised the data to draw it? How does the pie chart show the information? When do most people look at their phone according to this pie chart? What other type of graph could be used to show this data set?

B

Marcia is using this computer program to draw graphs for a presentation. What types of graphs can she draw? How do you think she chooses? What are the advantages of using a computer program to draw graphs?

Think, talk and write

A Reading and interpreting graphs *(pages 162–164)*

1 Look at the graphs used in this topic.
  a Name the types of graphs.
  b Discuss the data that each graph shows.

2 Look at this graph. Discuss why it is not very useful.

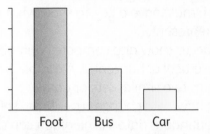

B Drawing graphs *(pages 165–166)*

1 Work in pairs. Look through some local newspapers or use the Internet to find examples of different graphs.
  a Describe the data each graph shows.
  b What features help you make sense of the graphs?
  c Is the graph correctly labelled and do the axes have a clear scale and regular intervals?

2 Computers and graphics packages have made it very easy for people to make graphs. Discuss how a computer graphics package works and what information the user has to provide.

C Probability *(page 167)*

1 Predict how likely these things are. Sort them into four categories: impossible, unlikely, likely, certain.
  a The Sun will rise tomorrow.
  b There will be snow outside your classroom tomorrow.
  c The next Olympics will take place in Grenada.
  d It will rain sometime tomorrow.
  e There will be six visible moons in the sky tonight.
  f You will go to bed after midnight tonight.
  g You will do some homework this afternoon.
  h One hour after 7 o'clock it will be 8 o'clock.

2 Add one more event of your own to each category.

Most dice are cubes. Have you seen any other kinds of dice? How many faces do these dice have? Compare the dice in the photograph. Do you still have an equal chance of rolling each number on the dice? Discuss your ideas in groups.

A Reading and interpreting graphs

Explain

Different types of graphs can be used to display different types of data.

* **Pictographs** use pictures or symbols to show and compare data that can be counted. They show how many or how much. A pictograph must have a **key** to show what each picture or symbol represents.
* **Bar graphs** use bars to show and compare countable data. The graph can be vertical or horizontal. A **double bar graph** shows and compares two similar sets of data.
* **Line graphs** show how data changes over time. This sort of data is called continuous data. A line graph can show patterns or trends in the data.
* **Pie charts** show how data is divided or shared. It shows how one part of the data is related to another part and also how it is related to the whole set of data.

Maths ideas

In this unit you will:
* read and interpret different types of graphs
* compare data sets and look for patterns on graphs.

Key words

pictographs	line graphs
key	pie charts
bar graphs	
double bar graph	

1   Study the pictograph and answer the questions.

Number of people swimming at Redoubt Beach at noon

Monday 12:00	🏊 🏊 🏊 🏊 🏊 🏊
Tuesday 12:00	🏊 🏊 🏊 🏊 🏊 🏊 🏊 🏊 🏊
Wednesday 12:00	🏊 🏊 🏊 🏊
Thursday 12:00	🏊 🏊
Friday 12:00	🏊 🏊 🏊
Saturday 12:00	🏊 🏊 🏊 🏊 🏊 🏊 🏊 🏊 🏊 🏊 🏊 🏊 🏊 🏊 🏊 🏊 🏊 🏊
Sunday 12:00	🏊 🏊 🏊 🏊 🏊 🏊 🏊 🏊 🏊 🏊

Key: 🏊 = 3 swimmers

a   What does this graph show?
b   What does one picture on the graph represent?
c   Which day had the most swimmers at 12:00? Do you need to count to answer this?
d   On which days were there only a few swimmers?
e   On which days were there the same numbers of swimmers?

2   The key on a pictograph is important as it tells you what each picture represents.

   a   If 😊 represents 150 people, how many people does 😊 😊 😊 represent?

   b   If 🌙 represents 300 bananas, how many does 🌙 🌙 🌙 show?

   c   ⬤ represents 400 oranges.

      i   How many oranges does ⬤⬤⬤⬤◖ represent?

      ii   Draw the symbols you would use to represent 2 200 oranges.

   d   🍬 represents 100 sweets. How many sweets are represented by the following?

      i   🍬 🍬 🍬   ii   🍬 ◖

3   This bar graph shows some of Pat's weekly expenses.

   a   How much does Pat save?
   b   How much rent does she pay?
   c   How much does she spend on food, clothing and entertainment altogether?
   d   How much will Pat spend on rent, food and clothing in four weeks?

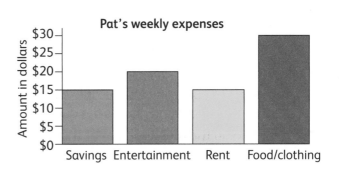

Pat's weekly expenses

4   This graph shows the number of followers that some well-known sport stars have on a popular social networking service. The numbers are in millions.

   a   Suggest a suitable title for this graph.
   b   Which sport star has the most followers?
   c   Approximately how many followers does Chris Froome have?
   d   What conclusions can you draw from the data? Discuss your ideas with a partner.

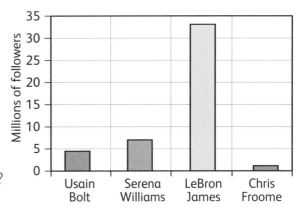

5   Study the double bar graph and answer the questions.

   a   What statistical data does this graph show?
   b   In which months was the monthly rainfall higher than average?
   c   In which month was the monthly rainfall lower than average?
   d   For what other types of data would a graph like this be useful?

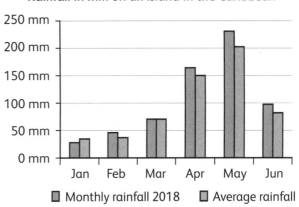

Rainfall in mm on an island in the Caribbean

■ Monthly rainfall 2018   ■ Average rainfall

6 This line graph shows the temperatures in Barbuda at 12 noon over a period of two weeks.

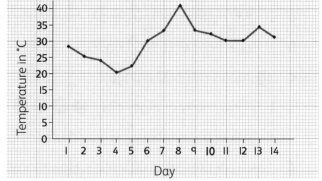

Daily temperature at 12:00 in Barbuda

a Which day had the highest temperature at noon?

b Which day had the lowest temperature?

c What were the highest and lowest temperatures?

d During which season do you think this graph was drawn?

e Between which two consecutive days was there the greatest drop in temperature? What was the difference in temperature between these two days?

f Between which two consecutive days was the greatest rise in temperature? What was the difference in temperature between these two days?

g How do you know if the temperature is rising or dropping on this graph?

7 A clothing shop used sales data to compile this graph of sales for the first six months of the year. The sales figures are in thousands of US dollars. Study the graph and discuss the questions.

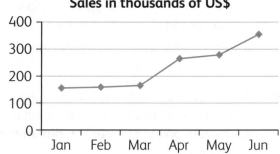

Sales in thousands of US$

a Should the management of this shop be pleased or unhappy?

b Why is a graph like this useful to shop management?

c Would it be useful to compare sales with sales at other times? How could that be done? What could be compared?

8 This pie chart shows how students in a class get to school each day. The pie chart has four sectors.

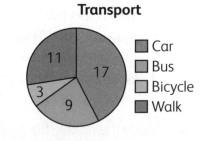

Transport

a How do you know what each sector represents?

b What is the total number of students represented in this data?

c How can fractions help you estimate what each sector of a pie chart represents?

d How else could you show this data?

What did you learn?

Answer these questions about the graph.

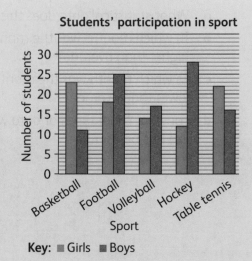

Students' participation in sport

1 Which sports have more boys participating than girls?

2 Which sport is the most popular among girls?

3 Which sport has the greatest difference in numbers of boys and girls participating?

4 How many students play basketball?

5 What do we call this type of graph and why is it useful for this data set?

B Drawing graphs

Graphs are useful for displaying information or data visually. Before you can draw a graph, you have to collect relevant data, organise it (using tally charts or frequency tables) and decide which type of graph would be best to represent the information.

When you draw a graph, you need to make sure that it has a heading and clearly labelled **axes**, or a **key** that tells you what symbols or colours represent.

These diagrams show the important features of bar and line graphs.

Maths ideas

In this unit you will:
* organise data and draw suitable graphs to represent it.

Key words

axes	scale
key	axis

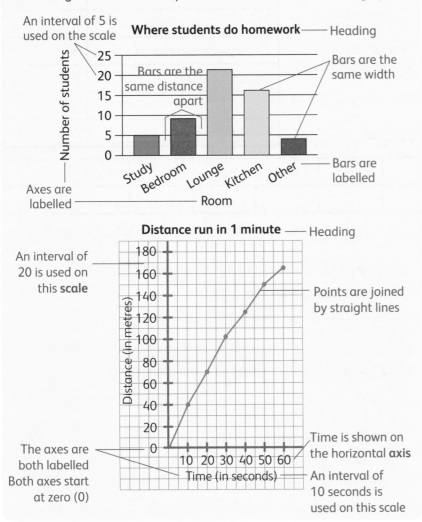

1 Which type of graph would you draw to display each of the following sets of data? Give reasons for your choices.

a Data that shows how many days of sunshine there are in Grenada each month of the year.

b Data that shows the time that people in different countries spend on social media each day.

c Data that shows the body temperature of a patient in a hospital every hour.

d Data that shows how much money a family spends on different items each month, as a percentage of what the family spends altogether.

2   Use the data in this table to draw a bar graph showing monthly rainfall. (Use graph paper to draw this graph. You can find blank graph paper on pages 97 to 100 of the Workbook.)

Month	Rainfall in mm
January	140
February	105
March	95
April	90
May	150
June	245

Now answer these questions about your graph.

a   Which month had the highest rainfall?

b   Which months had the same amount of rain?

c   How much rain fell during the first three months?

d   Which months had more rain than January?

e   How much rain fell for the six months?

f   What was the mean rainfall for the six months?

3   Tori wanted to find out which television programmes students in her school watched. She interviewed 30 students and recorded the data in this frequency table. Use the data to draw a graph. (You can find blank graph paper on pages 97 to 100 of the Workbook.)

TV programme	Frequency
Local news channels	5
International news channels	3
Sports channels	12
Series	6
Movies	4

4   The table shows the temperature of the water in the sea at different times during one day.

Time	9:00	10:00	11:00	12:00	13:00	14:00	15:00
Temperature (in °C)	21	22	24	26	28	29	29

a   Draw a line graph to represent this data. (You can find blank graph paper on pages 97 to 100 of the Workbook.)

b   Use the graph to estimate the temperature of the sea at 12:30 and at 14:30.

c   Comment on the temperature changes during the day.

Investigate

5   You will need: graph paper (from pages 97 to 100 of the Workbook), a ruler and a piece of string about 1 metre long.

What to do:

a   Ask a partner to measure your height in centimetres.

b   Measure your foot.

c   Use the string to measure the distance around your waist.

d   Use the string to measure the distance around your neck.

e   Record your measurements in a table.

f   Draw a graph to show the information in your table. Let each large square on the graph paper represent 10 cm.

What did you learn?

Write down three important things to remember when you are drawing a bar graph or a line graph.

C Probability

Probability tells you how **likely** something is to happen. If something is certain to happen, the probability is 1 or 100%. When there are several possible **outcomes**, you use probability to **predict** the likelihood of each outcome.

When you throw a normal six-sided dice, there are six **possible** outcomes: 1, 2, 3, 4, 5 and 6. You have an equal chance of each outcome.

The probability of throwing a 3 is 1 out of 6, or $\frac{1}{6}$.

We can write this as $P(3) = \frac{1}{6}$. P stands for Probability.

The probability of throwing any number that is not 3 (in other words 1, 2, 4, 5 or 6) is 5 out of 6 or $\frac{5}{6}$.

We can write this as $P(\text{not } 3) = \frac{5}{6}$.

$$\text{Probability of a favourable outcome} = \frac{\text{Number of favourable outcomes}}{\text{Total number of possible outcomes}}$$

An impossible outcome has a probability of zero or 0%. For example, when you throw an ordinary six-sided dice, there is a zero possibility of throwing a 7.

Maths ideas

In this unit you will:
* revise what you know about probability
* learn how to calculate probabilities.

Key words

probability	predict
likely	possible
outcomes	

1   Look at spinner A. Express as a fraction the probability of landing on:

    a red   **b** green   **c** any colour other than red or green.

Spinner A   Spinner B

2   Look at spinner B.

    a How many possible outcomes are there?

    b List the possible outcomes. Then express the probability of each as a fraction.

3   When you roll a fair six-sided dice, what is the probability of throwing:

    a a 6?   **b** an even number?   **c** an odd number?   **d** a number <7?

Problem-solving

4   Shaundra has a bag with 20 counters in four different colours: red, blue, yellow and purple. The probability of choosing each colour if she pulls out a counter without looking is given in the table. Use this information to work out how many of each colour counter there are in the bag.

Colour	Red	Blue	Yellow	Purple
Probability	25%	20%	10%	45%

What did you learn?

If there is a 9 out of 10 probability of rain today, what is the probability that it will not rain today?

Topic 15 Review

Key ideas and concepts

Write the mathematical word for each of the following and draw a rough labelled sketch to show the important features of each.

1   A graph that shows information for two related sets of data

2   A graph that uses pictures to represent data

3   A graph that uses equally spaced bars to show the values of data

4   A graph that shows how values change over time

Think, talk, write ...

How would you answer a student who asked these questions?

1   I know what data is, but what is data handling?

2   Why do we group tally marks in fives? Wouldn't it be simpler to make a long row of tallies?

3   I have to represent 16 oranges on a pictograph and one symbol represents 3 oranges. How do I show 16?

Quick check

1   Study the graph and answer the questions.

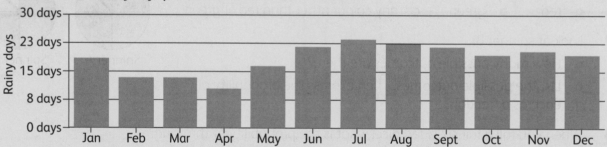

Mean number of rainy days per month in the Eastern Caribbean

a   How many rainy days are there on average in April?

b   Which three months have the most rainy days?

c   What is the difference in number between the most and the least days of rain per month?

d   How do you think the weather office collects and organises data to make graphs like this one?

2   Draw a bar graph to show how much time you spend sleeping, eating, attending school and doing other things each day. Label your graph clearly so other people can understand it.

3   Choose the correct words to complete each statement about this spinner correctly.

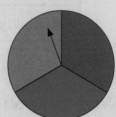

a   The arrow has a (1 in 3 / 1 in 6 / 50%) chance of landing on red.

b   The arrow is (more likely/less likely/equally likely) to land on grey as on red.

c   It is (possible/impossible/likely/unlikely) that the arrow will land on black.

Test yourself (3)

Explain

Complete this test to check that you have understood and can manage the work covered this year.
Revise any sections that you find difficult.

1   a   What is the missing number in the sequence?
        17, 23, 29, _____, 41
    b   Write the next three numbers in the sequence.
        2, 5, 11, 23, _____, _____, _____
    c   Explain how you found the next three numbers in sequence b.

2   a   What is the place value of the digit 9 in 49 328?
    b   In the number 4 583 201, which digit is in the ten thousands place?
    c   Write down a six-digit number and give the value of the third and fourth digits.
    d   Write eight thousand and eighty-five dollars in figures.

3   a   Which one of the following numbers is not a prime number?
        51   59   63
    b   Which one of the following is not a factor of 24?
        6   9
    c   Which of these statements is true for a set of whole numbers?
        i   Odd + even = even   ii   Even − odd = even   iii   Odd × even = even

4   a   Write 40 as a product of its prime factors.
    b   What is the lowest common multiple (LCM) of 6, 8 and 12?
    c   What is the highest common factor (HCF) of 8, 16 and 20?

5   Show how you solve each problem.
    a   A vendor wants to pack oranges into bags containing 4, 6 or 9. What is the least number of oranges he would need?
    b   What is the smallest number of sweets that can be put into bags each holding 3, 5 or 10 sweets?
    c   There are 85 fewer plums than cherries in a box. There are 327 cherries in the box. How many plums are there?

6   Jimmy has some marbles. He receives 9 more marbles. He now has 24 marbles. Write a number sentence that best represents this information.

7   a   The sum of two numbers is 3 458. If the smaller number is 1 567, what is the greater number?
    b   James has 47 marbles. Jim has twice as many as James. David has 13 fewer marbles than James. How many marbles do they have altogether?
    c   270 chairs are arranged in 6 rows. If each row contains the same number of chairs, how many chairs are in each row?
    d   Josh can do 6 haircuts in 4 hours. If he is paid $20 for each haircut, how many hours will it take him to earn $480?

8   Look at this diagram carefully.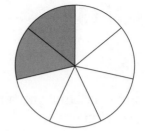

    a   What fraction of the circle is not shaded?

    b   Write the shaded amount of the circle as a decimal.

    c   Write these fractions in descending order:

$$\frac{2}{3} \qquad \frac{7}{12} \qquad \frac{5}{6}$$

9   Calculate.

    a $1\frac{4}{5} \times (\frac{5}{8} - \frac{1}{4})$ b $\frac{7}{8} - \frac{2}{5}$

10 a   Mrs Charles has $420. She gives $\frac{2}{3}$ of her money to her church. How much money does she have left?

    b   In a class of 24 students, $\frac{5}{6}$ like football. How many students like football?

    c   A water tank holds 360 litres of water when it is $\frac{3}{5}$ full. How many litres of water will the tank hold when it is filled to capacity?

    d   What length of ribbon would be left from a roll containing 1.2 m if a piece measuring 350 cm is cut off?

    e   A football pitch cannot be longer than 100 metres and cannot be wider than 70 metres. If this is the actual length and width, what is the perimeter of the football pitch?

11 What is the value of the 3 in the number 8.32?

12 a   Each passenger on an airline flight is allowed a bag that weighs no more than 25 kg. Johnny's bag weighs 31.3 kg. How much must he remove from the bag?

    b   Derek bought a notebook for $8.95 and a pencil case for $3.75. If he paid with a $20 bill, how much change did he receive?

    c   Find the sum of 7.5, 43.2 and 135.8.

    d   The sum of two numbers is 9.27. If the smaller number is 3.85, what is the greater number?

    e   Find the product of 4.95 and 2.1.

13 a   Express 0.7 kg in grams.

    b   At birth a baby has a mass of 3 600 g. What is its mass in kilograms?

    c   One book weighs 3 ounces. One bag weighs 25 ounces. What is the total mass of 5 books and the bag?

14 a   What is the base metric unit for how much a container holds?

    b   Sammy's juice bottle holds 3 litres of liquid. He drinks 750 ml at school and 500 ml when he arrives home. How many millilitres of juice remain in the bottle?

15 a   What is 20% of $120?

    b   What percentage of 200 is 40?

    c   There are 30 students in a room. Of these 60% are girls. How many girls are in the room?

    d   Ashanti scored 80 % in a mathematics test. The test was marked out of 40. What was Ashanti's score?

    e   If 20% of a number is 30, what is 50% of the same number?

    f   Jamie saved $120. Carolyn saved 25% less than Jamie. How much money did Carolyn save?

    g   Of the students in a class, 60% like to watch cartoons. The remaining 15 students like to watch sports. What is the total number of students in the class?

16 a An amount of money is shared among Jill, Sherwin and Angel in the ratio 1 : 2 : 3, respectively. If Sherwin receives $50, how much does Angel receive?

b On a farm there are 2 sheep for every cow. If there are 30 cows, how many sheep are there?

17 a What time is shown on the clock?

b If this time is in the afternoon, write it as it would appear in the 24-hour system.

c What would the time be 50 minutes later? Give your answer in the 12-hour system.

d What time is 20:15 in the 12-hour system?

e How many minutes are there between 2:05 p.m. and 3:30 p.m. on the same day?

f A bus travelled at an average speed of 90 km/h for $3\frac{1}{2}$ hours. What total distance did the bus travel?

18 1 cm on a map represents 25 km on the ground. What distance does 7 cm on the map represent?

19 Calculate the area of this rectangle.

9 cm
4 cm

20 The base of a right-angled triangle is 6 cm.
The height is twice its base.
What is the area of the triangle?

21 Look at the diagram on the right.

a Are lines AB and YZ perpendicular?

b Name two lines that are parallel in the diagram above.

22 Look at the diagram below.

a What is the size of the angle marked x?

b What type of angle is the 130° angle?

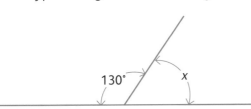
130° x

23 a Which one of the following shapes best represents an isosceles triangle?

A   B   C

b What is the size of angle a in this triangle? Why?

c What type of triangle is this?

d What do you call an angle that is equivalent to 180°?

e Debra says she drew a triangle with one right angle and two obtuse angles. Explain why her statement is false.

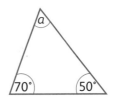
a
70°   50°

24 The ages of some students are given in the table.

Students	A	B	C	D	E	F	G	H
Ages	11	12	11	12	13	11	12	11

    a Give two ways in which this data could have been collected.

    b What age is the mode?

    c What is the mean age of this group of students?

    d Another student aged 13 is added to the data. How does this affect the mean age?

25 How many lines of symmetry are there in the shape below?

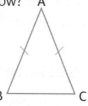

26 Name each type of angle.

27 The exchange rate for US$ to EC$ at a bank is US$1.00 = EC$2.60. If Mrs Mark exchanges US$150 for EC$, how many EC$ does she receive?

28 Which one of the following best represents the net of a cylinder?

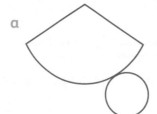
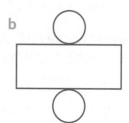

29 This graph shows the number of students who own pets in a class.

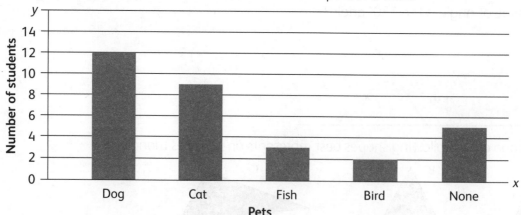

    a Write a suitable heading for the graph.

    b What interval is used on the y axis (the vertical axis)?

    c Which is the least popular pet among the students?

    d How many students own no pets?

    e How many students were surveyed when collecting this data set?